prenatal
life

prenatal life

biological and clinical perspectives

Proceedings of the Third Annual Symposium
on the Physiology and Pathology of
Human Reproduction

*compiled and edited by Harold C. Mack, M.D.
Wayne State University School of Medicine*

Wayne State University Press   Detroit   1970

Copyright © 1970 by Wayne State University Press,
Detroit, Michigan 48202. All rights are reserved.
No part of this book may be reproduced
without formal permission.

Published simultaneously in Canada
by The Copp Clark Publishing Company
517 Wellington Street, West
Toronto 2B, Canada

Library of Congress Catalog Card Number: 73-91873
Standard Book Number: 8143-1404X

contents

	Preface	7
	Introduction by Ernest Gardner, M.D.	9
I	Development and Differentiation of the Uterus *Emil Witschi*	11
II	Placental Circulation in Rhesus and Man *Elizabeth M. Ramsey*	37
III	The Physiologic Response to Uterine Contractions *Charles H. Hendricks*	55
IV	Adrenergic Mechanisms in Human Myometrial Control *Richard W. Stander and Tom P. Barden*	75
V	The Diagnosis and Treatment of Fetal Distress *Edward H. Hon*	91
VI	The Environment in which the Fetus Lives: Lessons Learned Since Barcroft *Donald H. Barron*	109
VII	Diagnostic and Prognostic Significance of Intrapartum Fetal Tachycardia and Type II Dips *R. Caldeyro-Barcia, A. A. Ibarra-Polo, L. Gulin, J. J. Poseiro, and C. Méndez-Bauer*	129
VIII	The "Ovariectomy Syndrome" and the Initiation of Labor *Arpad I. Csapo*	155
IX	Studies on the Human Fetus *Carl Wood*	183
X	Exogenous Factors Affecting Labor *Emanuel A. Friedman*	205

XI The Effect of Intrauterine Hypoxia/Asphyxia on the
Surviving Child   *Kenneth R. Niswander*   *217*
XII The Selection of Indicators of Fetal Circumstance
Karlis Adamsons   *235*
List of Contributors   *243*
Index   *245*

preface

In recent times the many gaps in our knowledge of the physiology and pathology of human reproduction have become matters of deep concern. On the one hand, we are today confronted throughout the world by unprecedented multiplication of the human species, often in catastrophic numbers, especially among underprivileged peoples. Excessive quantity of offspring is a matter of concern to economists, sociologists, and governments. On the other hand, even in the more privileged societies deficient quality of human offspring is also a matter of mounting solicitude. Waste and misery engendered by perinatal mortality and morbidity, as well as congenital malformations and defects, exact unconscionable tolls. The ultimate hope, of course, is that knowledge will permit control of both quantity and quality and thus prevent factors that now interfere with the rights of many infants to be physically and mentally "well-born."

In this quest for understanding of basic causes we must begin at the beginnings, namely, prenatal life, the subject of this symposium, and view it through both biological and clinical perspectives. "For the embryo in the uterus, pilgrim's progress begins with the process of attachment or placentation, by which the human child is to win his nine months of prefatory life. Thus early must he contend with his environment—which for the time being is the lining of his mother's uterus—and at the same time must adjust himself thereto." (George W. Corner, *Ourselves Unborn: An Embryologist's Essay on Man* [New Haven, 1944], p. 37.)

preface

The field of reproductive physiology and pathology is a many-sided one with intimate relations to many branches of biology and medicine. The interdisciplinary approach is exemplified by the distinguished authorities from several fields of science who participated in the symposium and contributed to this volume. A wide range of inquiries were considered, beginning with the embryology of the uterus and placentation and ending with a study of the effects of prenatal life upon the growing child. The intervening chapters detail some of the chemical, physical, endocrine, and exogenous factors affecting the fetus during its intrauterine existence and terminating in the still perplexing enigmatic chain of events that lead to the initiation of physiologic labor at term, and the pathologic physiology that culminates in prematurity. From reading these pages one can be reassured that some beginnings are being made in our explorations of prenatal life within the amniotic space. What we have learned and are learning since the days of Barcroft's pioneering efforts is presented here by Professor Donald Barron, who also points out some of the directions in which students of prenatal life must travel.

It is our hope that those who attended the conferences and those who read these pages will be stimulated to ask more questions than were answered. Thus only will these symposia realize their goal of focusing upon and enlightening the rapidly advancing frontier of the biology and medicine of human reproduction.

Harold C. Mack, M.D.

introduction

Wayne State University School of Medicine is collaborating with Harper Hospital in the presentation of symposia on the physiology and pathology of reproduction. In introducing the publication of the proceedings of this particular symposium, I think it pertinent to focus on certain topics that are enormously important to medicine and society.

These symposia have been arranged by and are being conducted by the Departments of Gynecology and Obstetrics in the school and at the hospital. Nevertheless, the disciplines that are represented include most of the biomedical sciences. The breadth of representation indicates the scope of the problems to be tackled. Consider, for example, the rapidity, precision, and autonomy of differentiation, beginning immediately after fertilization; the exquisite sensitivity of the young embryo to environmental and genetic damage at molecular levels; the development of integrated control of maturation and growth in the later fetal months, especially noticeable in the skeleton; the increasing awareness by the unborn child of the external world and the effect of that awareness upon the child; and finally the characteristics of this unique symbiosis and the ways in which preparation is made for severing the relationship. The challenge is this: To be able to offer to the unborn child the kind of medical and health care and protection that he deserves and has a right to expect. This challenge cannot be met without a significant, in fact, exponential expansion of research, ranging from bio-

introduction

logical research that uses lower animal forms in an effort to solve problems that are difficult or impossible to tackle experimentally in placental animals, to psychological and sociological research that deals with the impact of social, cultural, and economic factors on the unborn child. What will develop, I am convinced, is an obstetrics that is much more broadly based, that pays much more attention to, indeed, specializes in, the sequence and problems of prenatal development, and that ultimately will be able to treat or prevent many of the now disabling or lethal prenatal events.

It cannot be stressed strongly enough that the biological and medical features of these problems, and their study and solution, cannot be divorced from social, cultural, and economic aspects. These present the world today with some of its most serious problems. Consider the staggering immensity of overpopulation and the almost insurmountable roadblocks to its management posed by religion, history, culture, level of education, and politics. Consider our shameful record in maternal and infant mortality in the United States, especially in the poverty-stricken parts of our cities. Consider the degree of ignorance, even among physicians, about sex, reproduction, and prenatal life, and reflect that many think the answer is to leave it all to a high school teacher.

Finally, it must be emphasized that the disciplines which are represented at this symposium are those that make possible, indeed, made necessary, the current emphasis on the problems of aging and of the chronic and degenerative disorders. The complexity and cost of medical health care rise in proportion to, perhaps out of proportion to, our success in treating and preventing the ills of younger years. It is unfortunate that the two ends of the life span are often considered separately in terms of processes, and particularly so in terms of national priorities for financial support.

This symposium, then, is an index of the biological bases of medicine and of the social environment in which medicine is practiced. Finally, it is an index of the future, one of brighter hope for the unborn child.

Ernest Gardner, M.D., Dean
Wayne State University School of Medicine

chapter I

Development and Differentiation of the Uterus

Emil Witschi, Ph.D., M.D.

During the last one hundred years an extensive literature dealing with gonaduct development in man has accumulated. Changing interpretations and application of new research methods together with the ever-live interest in all matters bearing on human reproduction will certainly keep production flowing in the foreseeable future.

The Primitive Gonaducts

Three primordial sources contribute to the development of the genital ducts: the urogenital sinus, the mesonephric ducts, and the oviducts.[a] Their embryonic history starts at four weeks[b] with the organization of pronephric tubules in the lowest of the cervical nephrotomes. Connecting by their lateral ends they give origin to paired nephric ducts. The short upper segments regress again early, while the sprouts growing caudally along the mesonephric blastema cords assume, as mesonephric ducts, important roles in the gonaduct development of both sexes. Their tips reach the left and right lateral walls of the cloaca within about two days.[c] When later the urorectal septum divides

---

[a] The popular name *oviduct* will be used alternatively with the technical term *paramesonephric duct*. "Length" for embryos usually is "greatest length" and for fetuses it is "crown-rump length." Ages are estimated on the basis of available data.

[b] Embryo stage 17; length 3.5 mm; 24 somites; age twenty-eight days.

[c] Embryo stage 19; length 4.0 mm; 29 somites; age thirty days.

prenatal life

the cloaca frontally, the mesonephric ducts maintain attachment to the ventral part, i.e., the future urogenital sinus. They carry with them short metanephric ducts.[d] Presently follows the absorption of the common terminal limbs into the dorsal wall of the urogenital sinus and the consequent formation of two pairs of orifices. The metanephric ones move distally, toward the bladder, whereby a considerable piece of mesoderm becomes inserted into the otherwise endodermal epithelium of sinus and urethra[e] (see Witschi,[16] pp. 527-531, for reconstructions of these stages).

The oviducts develop only after the mesonephric ducts have become attached to the urogenital sinus. In forty-day embryos[f] at the level of the third thoracic somite and directly below the pleuroperitoneal fold appears a patch of thick epithelium over the top of each mesonephric body. These funnel fields crease and fold, and from their grooves several nephrostomal depressions penetrate the subjacent mesenchyme. Some make connections with rudimentary kidney tubules, and on each side one becomes the funnel of an oviduct. By terminal growth these tubes elongate, first following the mesonephric ducts at some distance. Then their tips press against the lateral walls of the mesonephric ducts and get implanted into them. This relationship prevails during the entire downgrowth of the oviducts. The tips always retain an attachment; but as they move and proliferate, the new lengths of oviducts progressively separate from the mesonephric ducts (Fig. 1). At their pelvic ends the union is more lasting, at least in female development, as the following will show.

Even though sexual differentiation of the gonads has become recognizable in embryos of 17 mm length and an age of seven weeks, for a while the gonaducts still continue developing along identical lines in both sexes. During another week the oviducts complete their growth toward the urogenital sinus (U.G. sinus). In an embryo of two months[g] they run separately down the genital cord (Fig. 2), each one by its end fused to the mesonephric duct at its lateral side (Fig. 3). Already the tips of the oviducts start uniting also among themselves. By the end of one more week a single oviduct, the uterine canal, has become established in males (Fig. 7) as well as in females.

[d] Embryo stage 26; length 8.0 mm; 38 somites; age thirty-eight days.
[e] Embryo stage 30 to 32; length 14.6 to 19 mm; age forty-six to fifty-two days.
[f] Embryo stage 27; length 10 mm; age forty days.
[g] Embryo stage 33; length 25 mm; age fifty-eight days.

development of the uterus

So far attention has been given only to the epithelial primary mesonephric and paramesonephric tubes. However, both constantly attract numbers of mesenchymal cells, thus investing themselves with heavy coats of round and fibroid cells, which also will play important roles in uterine development. The upper oviducts in addition are wrapped in suspensory peritoneal folds. Approaching the pelvis these folds swing mesially and, carrying with

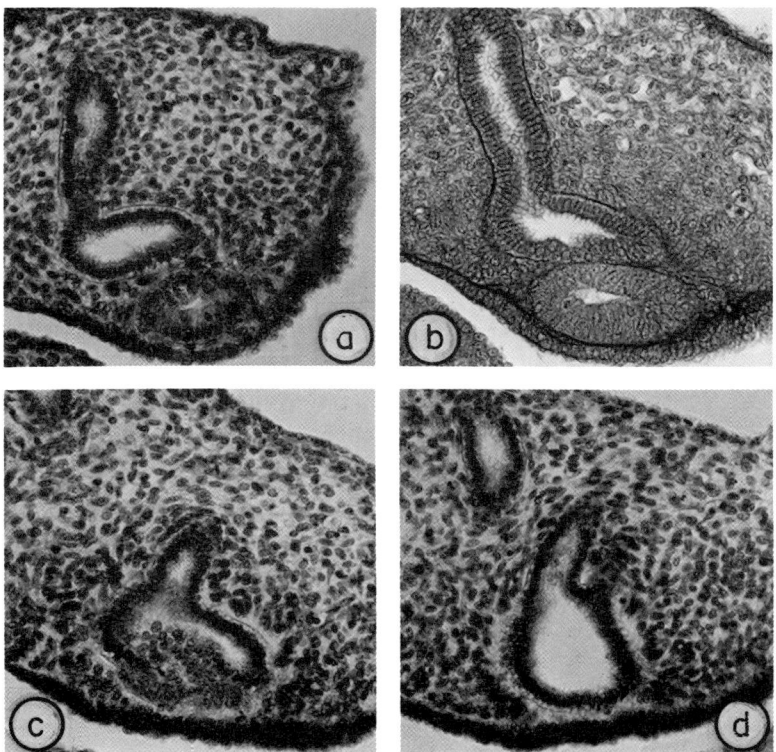

Figure I-1. Embryo 19 mm.[e] Succession of sections through mesonephros showing implantation of oviduct in lateral wall of mesonephric duct. a) In upper mesonephros the oviduct at some distance of mesonephric duct. b) The two ducts juxtaposed. c) Blastemic tip of oviduct fused with lateral wall of mesonephric duct. d) Below junction, single mesonephric duct. ×200.

13

prenatal life

them also the mesonephric ducts, left and right unite in the genital cord. Here the mesenchymal condensation around the ducts becomes very massive. The oviducal suspensories—later the broad ligaments—serve as bridges for the further supply of the genital cord, particularly with blood vessels and nerve

Figure I-2. Embryo 25 mm.[g] Cross section through urethra and lower genital cord, where oviducts are closely apposed; mesonephric ducts and oviducts imbedded in dense mesenchyme of genital cord. Separate urethral cord. Uppermost right side ganglia of pelvic plexus. ×67.

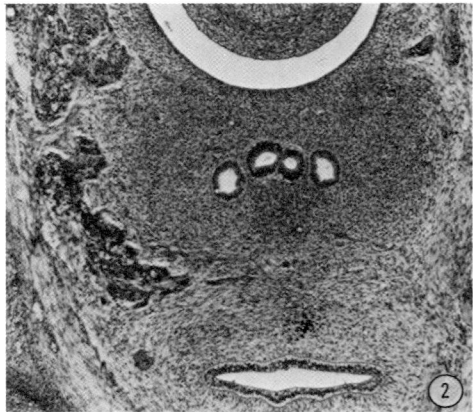

Figure I-3. Same embryo, section at lower level. Genital and urethral cords fused. At right, terminal tubule of mesonephric duct with orifice in urogenital sinus (U.G. sinus); at left, slightly higher level, mesonephric duct close to oviduct, terminal tubule bending out toward U.G. sinus. Walls of both ducts in partial desaggregation, fusion. Left and right pelvic ganglia plexuses. ×67.

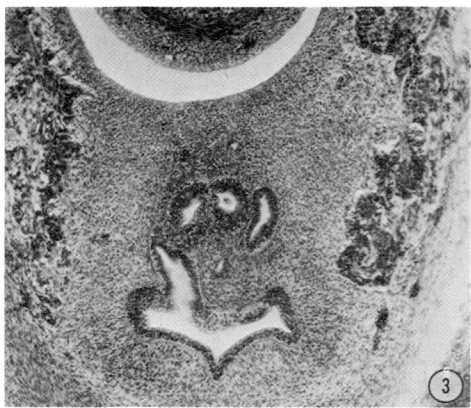

development of the uterus

bundles. The latter branch off from the early developing pelvic plexuses of sympathetic ganglia (Figs. 2, 3). Mesenchymal accumulation is particularly dense on the ventral and dorsal surfaces of the lower uterine canal.

A sagittal section through a slightly older fetus[h] (Fig. 4) shows the conditions existing in both sexes immediately preceding sexual differentiation of the ducts. The single oviduct canal extends from the top of the genital

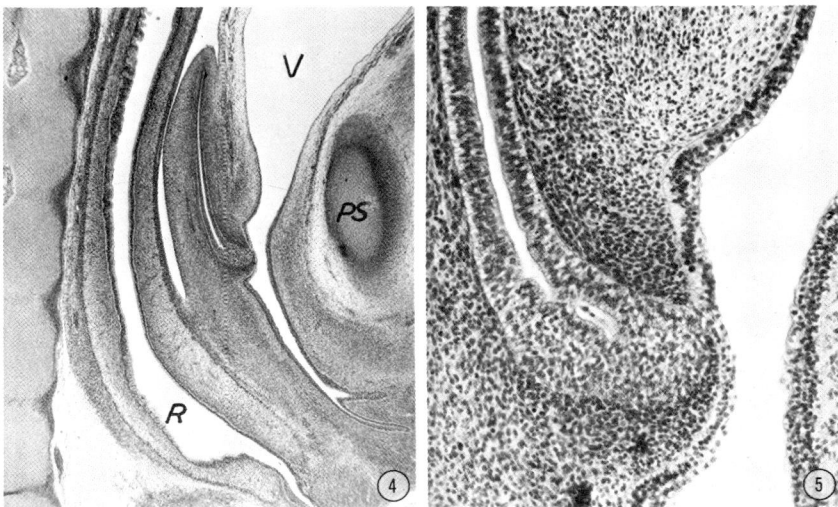

Figure I-4. Fetus 37 mm.[h] Part of sagittal section: last lumbar and five sacral vertebrae at left, pubic symphysis at right. P.S. pubic symphysis, R. rectum, V. vesica-urethra—U.G. sinus—vestibule. Genital cord adheres to dorsal wall of bladder, not to rectum. Uterine canal ends in large blastemic tip, projecting toward U.G. sinus. ×23.

Figure I-5. Same fetus. Detail of swollen tip of oviduct (uterine rudiment) separated from wall of U.G. sinus by wedge of mesenchyme from below, to where it meets edge of ostium of left mesonephric duct (mostly in following sections). Ostial epithelium continues upward in direction of bladder. ×127.

[h] Fetus stage 34; length 37 mm; age sixty-three days.

prenatal life

cord (level of the fifth lumbar vertebra) into the papilla of the orifices of the mesonephric ducts. The latter projects into the urethra behind the lower pubic symphysis (level of the second sacral vertebra). The end of the oviduct now is a compact mass of cuboid cells fused with the mesial walls of both mesonephric ducts. In the midsagittal section (Fig. 5) only a small edge of the left mesonephric duct becomes visible above the end of the oviduct. The protruding meristem bulb is capped by mesenchyme and covered by sinus epithelium. Its distance from the vestibule equals the length of the unpaired oviduct. Graphically the essential relationships are indicated in Figure 6A.

Sexual Differentiation of the Gonaducts

Significantly, the start of somatic sexual differentiation coincides in time with a spurt of massive increase in the interstitial component of the testes of male embryos followed by assumption of luteal character of its cells.[18]

In embryos of 30 mm the genital ducts between the lower ends of the mesonephric bodies and the genital cord (the so-called horizontal segments) are first to react to presence or absence of testes. In females the mesonephric ducts shrink while the oviducts continue growing and folding. In males, to the contrary, the oviducts become thin, and resorption of their epithelial lining follows almost immediately.

Regression of heterologous ducts proceeds slowly in the mesonephric and gonadal region and remains incomplete. Of particular interest, however, are the later developments in the genital cord and the urogenital region.

Male differentiation is a relatively simple process. The mesonephric ducts—now the deferent ducts—retain their original points of attachment at the urethra (compare Fig. 6, graphs A and H). This region becomes well marked by the rapid development of many prostate buds and tubules, below and above the orifices (Fig. 7). The development of this prostatic urethra obviously is stimulated by testicular secretions. In this fetus of three months[i] all essential characteristics of male differentiation are well laid out. The epithelial uterine canal has disappeared from the genital cord, except for its lowest part, which becomes the vestigial male utricle (Fig. 7). It still is connected with the deferent ducts by strands of darkly staining cells (Fig. 8).

Below the prostatic urethra the urogenital sinus is now set off as a

---

[i] Fetus stage 35b; length 75 mm; age three months, male.

development of the uterus

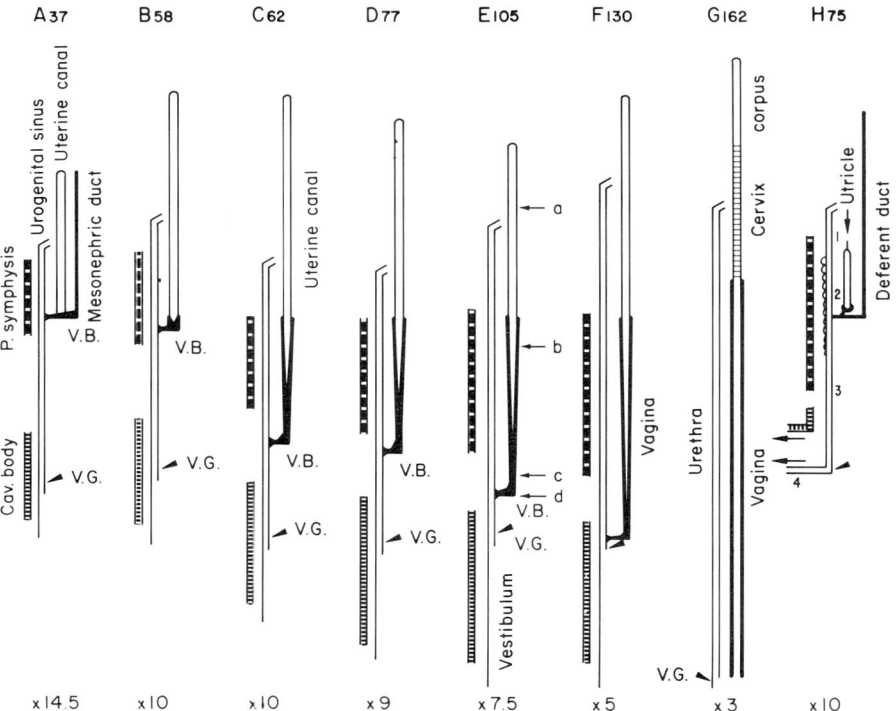

Figure I-6. Graphic presentation of longitudinal dimensions and relative arrangements of parts of urogenital organs and the pubic symphyses of eight fetuses (A to H). The figure following each letter indicates crown-rump length. On the bottom line, the enlargement used for each graph; the broken upper end of the urethral tube (U.G. sinus) indicates as closely as possible the level of the entrance of the ureters into the bladder. V.B. vagina bud, V.G. level of orifice major vestibular gland. Arrows along the right of graph E indicate level of sections shown in Figures 10-13. In graph G, the vestibule is omitted and the vagina-cervix limit is down at higher level for only technical reasons. A to G are females, H a male. 1-4, urethra.

membranous urethra. At the level of the vestibule and beyond the bulbo-urethral glands the passage becomes narrowed through formation of the raphe. Running between and below the cavernous bodies, the urethra turns off ventrally at a right angle and elongates considerably, to end in the glans of the penis (Fig. 6H).

prenatal life

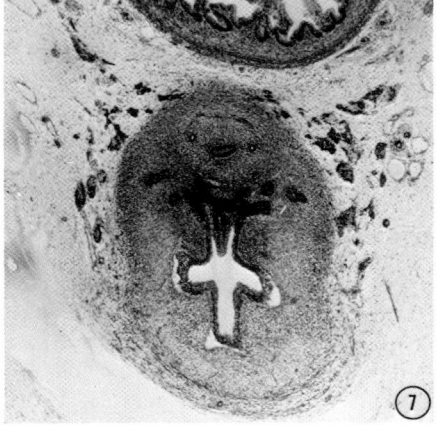

Figure I-7. Male fetus 75 mm.[1] Cross section, prostatic urethra and (enclosed) lower genital cord with mesonephric ducts and masculine utricle. ×67. (Compare Fig. 6, graph H.)

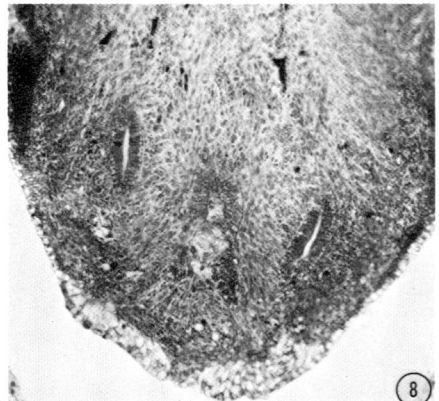

Figure I-8. Same fetus. Papilla of mesonephric (ejaculatory) ducts. Low end of utricle attached to patches of ostial epithelia by pair of mesonephric duct-blastema cords and fibrous strands. Deferent ducts a few sections above orifices. ×130.

In female differentiation the uterine rudiment becomes supplemented by the vagina, which at the end of its growth reaches the vestibulum (Fig. 6, compare graphs A and F). Throughout its descent the attachment to the sinus by the pair of terminal mesonephric tubules is maintained (Fig. 6, B to F). Growth is by proliferation from the meristemic vagina bud (V.B. in graphs) that becomes organized at the low end of the uterine rudiment. The center is

development of the uterus

furnished by the solid tip of the oviduct; its lateral parts by fused-on mesonephric duct blastema. With progress of the descent the contribution of the latter relatively increases while that of the oviducal part decreases.

The progression of this remarkable development is represented graphically with selected stages in Figure 6. Each graph is based on measurements from a single fetus. The relative positions of parts are determined by distances from the top of the genital cords and the openings of the vestibular glands. Cavernous bodies and roof of the vestibulum, ending with the clitoris, take a more or less slanting direction in female specimens, but this is not represented in the diagrams. All parts are arranged as in left lateral views; only the vagina rudiment is given as if sectioned frontally. Only longitudinal dimensions are true to scale. Their absolute values may be obtained with the help of the enlargement factors at the base of each graph. Descriptions of all cases are following hereafter.

a) At two months the fetus[h] exhibits the fully developed indifferent morphology as already described above. The mesonephric ducts open with flaring, bell-shaped ostia into the urogenital duct, only a short distance below the ureter attachments to the neck of the bladder, behind the pubic symphysis, and a considerable stretch (1.5 mm) above the vestibular (bulbourethral) glands (Fig. 4). The lower tip of the oviduct is tightly squeezed between and fused to the mesonephric ducts.

b) Half a month later in the 58 mm fetus[j] that serves as the basis for graph B (Fig. 6) the vagina bud (V.B.) has initiated growth and differentiation. A short tube, the first link of vagina, has been formed, immediately below the uterus rudiment, and in continuation of it. It consists of dorsal and ventral oviduct lamellae that are flanked by compact bodies (short pillars) of mesonephric duct blastema. In the section shown in Figure 9a, the knife cut slightly through the terminal tubule that connects the right body with the U.G. sinus. On the left (right in the picture) this connection lies in next following sections. A dense accumulation of mesenchyme separates the vagina from the sinus. In section 9b, 20 μ deeper, the right tubule contains some secretion. The left tubule points toward the U.G. sinus, but does not cut entirely through the mesenchyme strand passing to the central knob. The vaginal cavity is disappearing. Section 9c, again 10 μ more caudally,

---

[j] Fetus stage 35a; length 58 mm; age two and one-half months.

prenatal life

shows a much reduced round and solid tip of oviduct blastema enclosed by more prominent clasps of mesonephric duct blastema. The knob of dark staining mesenchyme is reaching down to this level. Section 9d, passing through the very bottom of the cup-shaped vagina bud, shows the lowest tip of the oviducal component and also parts of the terminal tubules of the mesonephric ducts.

c) Extensive growth caudally follows immediately the establishment of

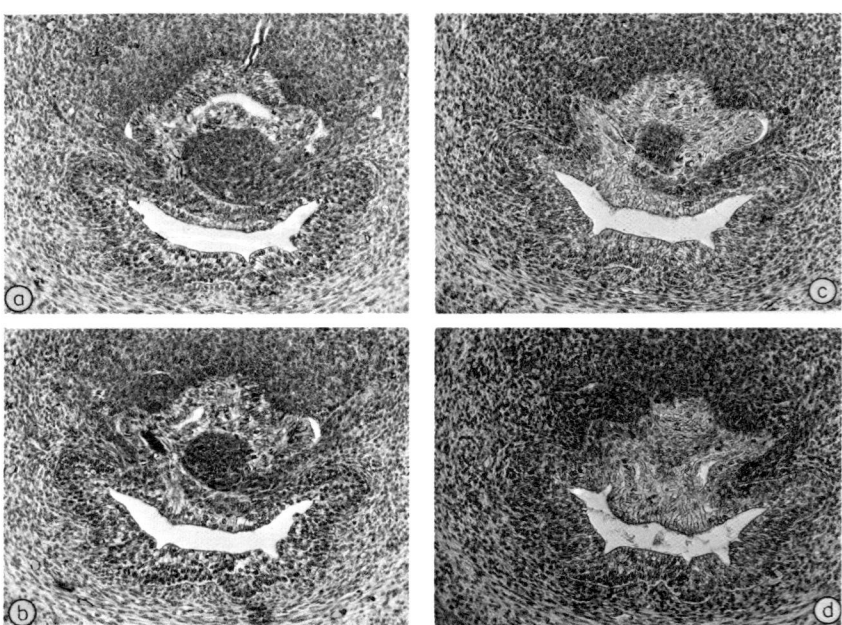

Figure I-9. Fetus 58 mm.[j] Cross sections through early vagina bud (see Fig. 6, graph B). a) Vagina consisting of flat oviductal tube and lateral mesonephric duct blastema-bodies. Edge of right terminal tubule connects with U.G. sinus. b) 20μ more caudally the right tubule with secretion in lumen; the left tubule is kept separated from sinus by mesenchymal strand. c) 10μ further down, both tubules join U.G. sinus, enclose plug of mesemchyme. Tip of oviduct blastema between mesial processes of mesonephric duct blastemas. d) Again 20μ further down; dorsally, end of oviduct part of the bud; laterally, terminal tubules of mesonephric duct component and high cylindric ostial epithelium protruding into U.G. sinus. All figures ×73.

development of the uterus

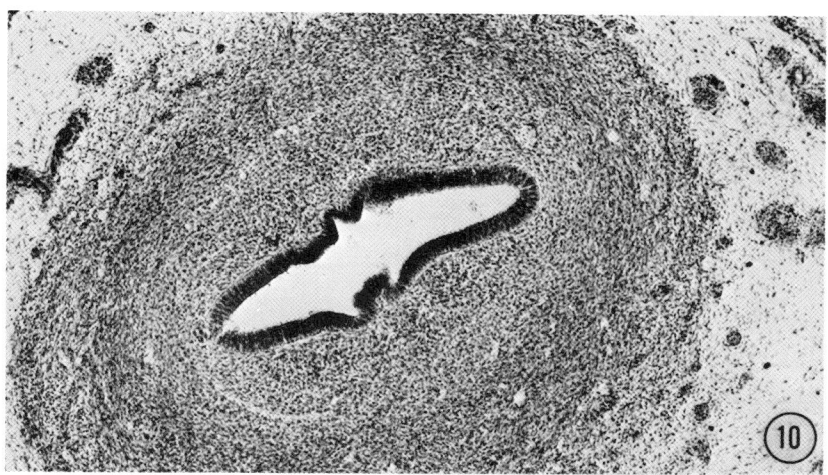

Figure I-10. Fetus 105 mm.ᵐ Cross section through genital cord (level a, graph E). The smooth cylinder epithelium of the uterine canal shows folds where the two primitive oviducts were united. ×75. (Carnegie collection No. 1908)

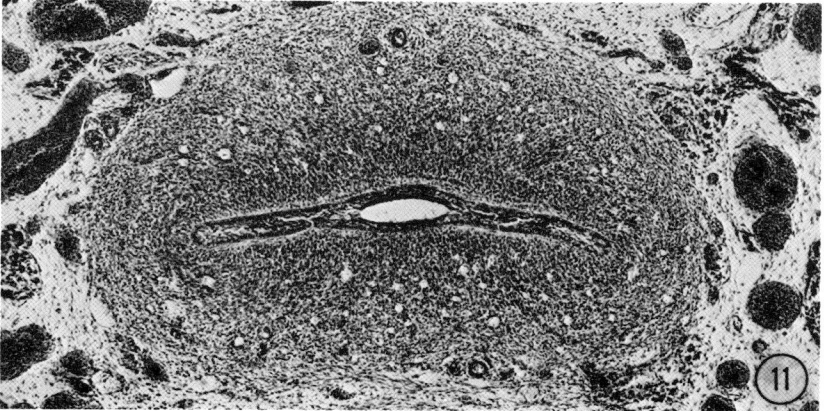

Figure I-11. Same fetus. Cross section through upper vaginal rudiment (level b, graph E). In center oviduct component with lumen dorsal and ventral cuboid epithelia. Long left and right fins of irregularly arranged cells of mesonephric duct blastema. Mesenchyme richly vascularized. ×75.

prenatal life

the composite bud. In a but-slightly-larger fetus[k] parallel contributions of oviduct and mesonephric duct blastemas result in prolongation of the central flat tube and formation of lateral fins. The tip has now passed the lower edge of the pubic symphysis and moves toward the vestibulum (Figs. 6b, c).

d) The same trends are maintained by a fetus of three months (77 mm).[l] For the understanding of the special characteristics of female differentiation, comparison with the male of similar size and age[i] should be enlightening (Fig. 6, graphs D and H). The male urethra consists of vesical, prostatic, membranous, and cavernous divisions. In the female similar sequences of segments cannot be distinguished. The most significant difference is the fact that in the male the orifices of the mesonephric ducts maintain their location in the prostatic urethra, i.e., at the level held from before the start of sexual differentiation (Figs. 3, 4), while in the female they are sliding down toward the vestibulum. Of the uterine canal only a vestige, the male utricle persists, while in the female an accessory vagina rudiment enlarges this sex duct. The meristemic bud, which developed before sexual differentiation, is vigorously proliferating in the female but has regressed to unimportant remnants in the male (Fig. 8).

e) In the now following fourth month the vaginal canal in its continued downgrowth approaches the vestibule. In a fetus slightly over 10cm,[m] cross sections at four levels (marked by arrows a, b, c, d in Fig. 6, graph E) show the purely oviducal lining of the uterine canal in the upper genital cord (Fig. 10), the addition of lateral wings originating from mesonephric duct materials to the vaginal canal (Fig. 11), and the terminal almost purely blastemic branching lamellas of the lowest part of the vaginal rudiment (Fig. 12). At the very end, connection with the urogenital sinus is established through the pair of ostia of the mesonephric ducts (Fig. 13). Even though it can not be definitely excluded that stray oviducal cells may become included also in the terminal lamella, evidently the organized tube, lined with cuboid or cylindric oviduct epithelium, constantly narrows and ends 500 μ above the ostial attachment to the sinus.

f) Reaching the level of the vestibular glands toward the end of the

---

[k] Fetus stage 35a; length 62 mm; age two and one-half months.
[l] Fetus stage 35b; length 77 mm; age three months.
[m] Fetus stage 35b; length 105 mm; age three and one-half months, Carnegie collection No. 1908.

development of the uterus

fourth month[n] the vaginal rudiment gains contact with the vestibulum. At this stage, sagittal sections through the full length of the system (Figs. 14, 15, and 16) still show the attachment of the lower end to the urethra by mesonephric duct elements.

In length the vaginal downgrowth of the last six weeks, from the uterine rudiment to the vestibule, equals now that of the uterine rudiment (Fig. 6,

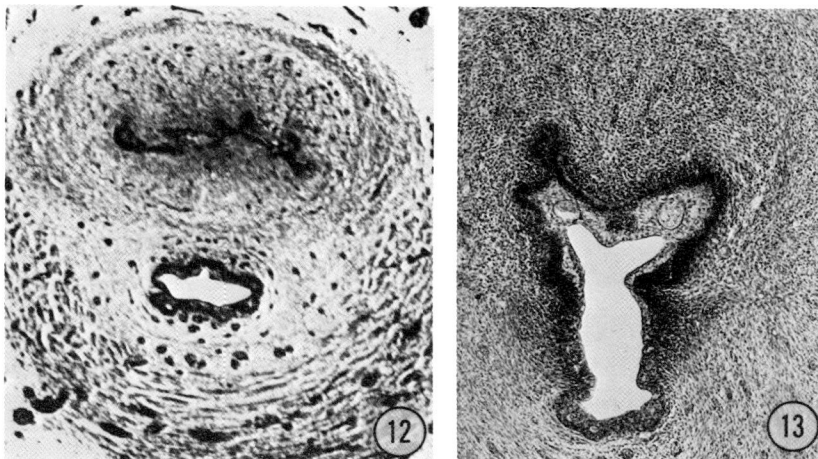

Figure I-12. Same fetus. Cross section through lower vaginal rudiment (level c, graph E). No organized oviduct component, duct blastema forms lobate, thin sheet. ×40.

Figure I-13. Same fetus. Cross section through tip of vaginal rudiment. Terminal tubules of the mesonephric ducts, opening into U.G. sinus. ×75.

graph F). The epithelium of the uterus is entirely of oviduct origin; that of the new addition, the vagina, is contributed in almost equal measures by oviduct and mesonephric duct proliferations, or occasionally incomplete

[n] Fetus stage 35c; length 130 mm; age four months, Carnegie collection No. 1575.

prenatal life

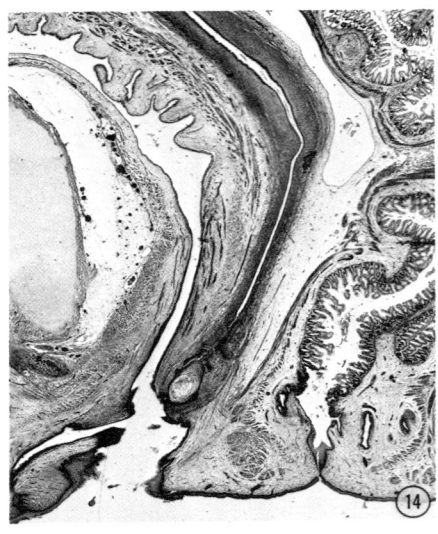

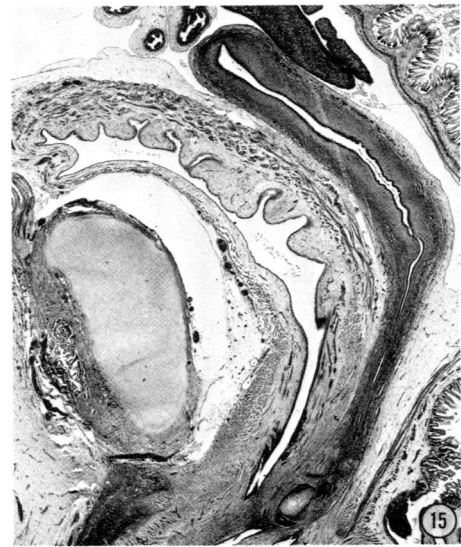

Figure I-14 and 15. Fetus 130 mm.[n] Sagittal and parasagittal sections through vesica, urethra, vestibulum, uterus, cervix, vagina, lower intestines, and anus. Lowest part of vagina inflated. ×6.5. (Carnegie collection No. 1575)

development of the uterus

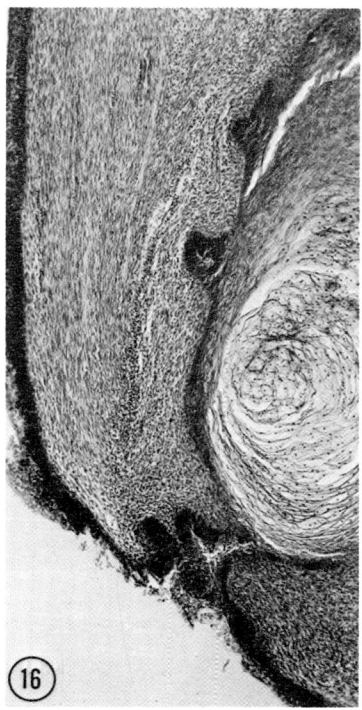

Figure I-16. Same fetus. More enlarged part of lowest vagina Figure 14. Terminal tubule and orifice of mesonephric duct, connecting vagina with urethra and vestibule. Prekeratinization of vaginal epithelium. ×66.

septa. Where left and right components meet, the anterior and posterior vaginal columns develop longitudinal ridges. The junction of the vagina to the uterine rudiment is marked by a circular groove, the fornices. This will hold true in a gross anatomical way, even though later the lining epithelia of vagina and cervix may not meet in the groove but rather at the top of the os or even inside the canal of the portio vaginalis of the cervix.

Functional Development of Major Segments of the Female Gonaducts—Inception of Estrogen Responsiveness

When the uterovaginal tract has become organized in its full length, that is at the fetal age of four months, the maternal and fetal bodies are flooded by increasingly high concentrations of estrogenic steroids.

prenatal life

According to Schwers,[14] the estriol content of maternal plasma increases from 1.8 μg/100 ml at four months to 16 μg/100 ml at ten months and the urinary excretion from 8,000 μg/24 h at five months to 30,000 at ten months of pregnancy. Other estrogens are present only in minor proportions. In fetal blood and plasma, estrogens are about ten times higher than in the maternal. Responsiveness of the fetal organs develops only gradually and—surprisingly—it starts with the last formed, low end of the vagina.

f) (continued) The sagittal sections (Figs. 14, 15) of the 130 mm embryo[n] show inflation and prekeratinization of the lower half of that part of the vagina which seems to originate purely from mesonephric duct blastema. This is the first distinct estrogen response. Some uninflated appendices hang limp on the surface. The short orifice tubules, which had conducted the descent of the vaginal bud, still pass through the mesenchymal sheath and connect with the urethral wall where this now widens in the vestibulum (Fig. 16). Above the small balloon a solid stretch of similar blastemic material is still unaffected, and the much longer upper vaginal canal is lined only by unstimulated oviducal epithelium.

In sections the mesenchymal sheaths appear thinner in the vaginal region than around the uterocervical canal. The cervix is not sharply separated from the corpus, but at least two-thirds of the entire uterus are set off by a narrow isthmus against the short, bulbous corpus. No indication of hormonal stimulation is noted in the corpus. In the massive mesenchymal cover some layering presages differentiation of mucosa, muscularis, and serosa.

g) Within half a month in a 162 mm fetus[o] estrogen sensitivity has spread over the entire length of the vagina (Fig. 17). The epithelium has become many layered. From the lively proliferating cuboid basal layer develops a thick squamous epithelium. It is now richly waved and through ingrowth of mesenchyme between the folds obtains a papillary appearance. This development extends up to the entrance of the cervix. The vagina has gained much in relative length (Fig. 6 G).

With the excessive development of the vagina, limits between its oviduct and deferent duct derived parts seem to have disappeared. Also the former ostia can no longer be located. Possibly the tubules have been absorbed into the walls of the vagina. Coincidentally, it is agreed by most ob-

---

[o] Fetus stage 35c; length 162 mm; age four and one-half months, Carnegie collection No. 5768.

development of the uterus

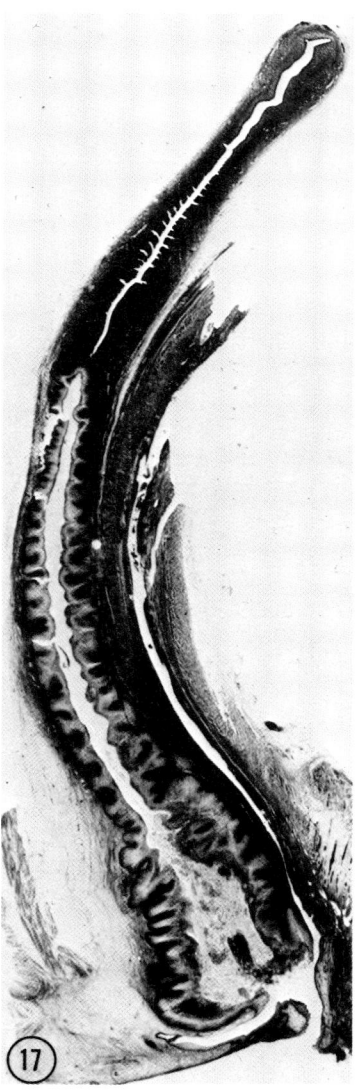

Figure I-17. Fetus 162 mm.⁰ Sagittal section showing clear differentiation of corpus, cervix, vagina, hymen, hymeneal pore. ×5.5. (Carnegie collection No. 5768)

prenatal life

servers that perforation of the hymen, i.e., of the thin wall between vaginal and vestibular cavities, usually occurs in fetuses of about this stage. Although I have not made detailed studies on this question a comparison of the 130 mm and 162 mm fetuses and of Figures 14, 16, and 17 suggests to me that the fetal hymeneal "perforation" may arise simply by reopening the old orifices of the mesonephric ducts. Incorporation of their walls may contribute to the formation of the hymen. Dehiscence of the ostia would be caused by pressure developed in the inflated vagina.

The epithelium of the endocervix is only slightly stimulated. It has changed from cuboidal to cylindrical cells and mucoid development starts in the grooves of the many narrow folds (Fig. 17).

In the mesenchymal coat of the uterus appear now unstriated muscle cells, but the endometrial part of the corpus remains still unchanged.

Gonaduct Development during the Last Six Months of Fetal Life

After the fourth month the major changes still occurring until birth concern the growth and the histology of vagina, uterus, and tubes. Even before the era of steroid chemistry, observation of considerable growth of the fetal sex organs during pregnancy and of their abrupt reduction following birth had suggested the existence of hormonal relationships between fetus and placenta. Halban[7] and Scammon[13] are among those who prepared the ground for modern sophisticated investigations on physiology and endocrinology of the feto-placental unit.

Hunter[8] has summarized in some telling graphs the growth of the female generative organs. Connecting the weight of the uterus at the stage of beginning hormonal responses (five months) with that after birth by a straight dash line, he concludes that the excess development is to be credited to hormonal stimulation (Fig. 18). There persists a remarkable preponderance in size and growth of the cervix over the corpus of the uterus. By the time of birth the former has attained more than twice the length and many times the weight of the latter. Even after postnatal involutioin the corpus still remains small in comparison to the size of the cervix. Adult proportions are established only at the time of puberty. However, during the prenatal six months the uterus gains its final shape, mainly by absorbing tubal material and acquiring the convex rounded top of the fundus.

development of the uterus

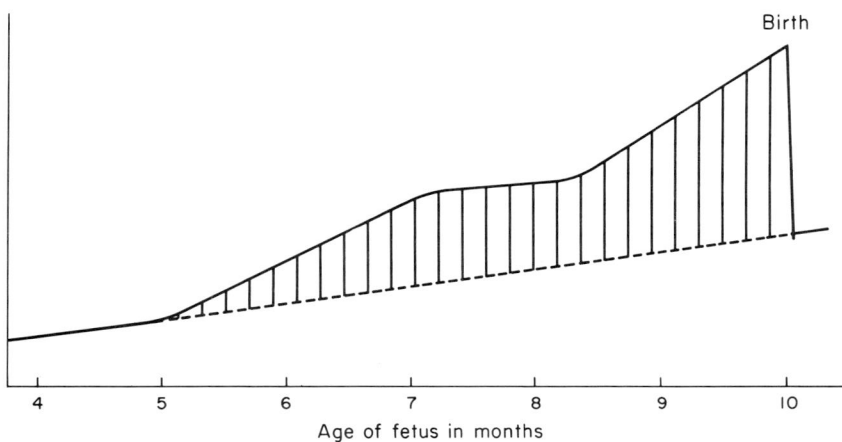

Figure I-18. Diagram (redrawn after Hunter[8]) suggesting separate embryonic and endocrine components of uterine growth during pregnancy.

The growth of the tubes is also greatly stimulated by the placental hormones. Following the sixth month the rich folds and lamellas form within, while in toto the tubes become folded into many loops. Having thus completed their embryonic differentiation, they regress less than the other parts, after birth. A comprehensive study of the histologic development of the ovaries and all parts of the genital tracts of female fetuses has been made by Rosa.[12] Starting with fetuses of the age of four and one-half months he recognizes a gradient along the genital tract in regard to first appearance and degree of estrogen sensitivity. The maximum lies in the vaginal epithelium and the lowest value at the fundus of the uterus. At the beginning the histologic reactions in all organs are purely of estrogenic type; but later a luteal effect becomes added. The second phase does not start simultaneously in all organs, depending probably on time differences in attainment of responsiveness and on dose requirements. The sequence of histologic reactions must quite naturally resemble that of a menstrual cycle. In fact for the endometrial epithelium and stroma, Rosa points out the following near-correspondences: 6 months (sixth menstrual day); 7 months (eleventh menstrual day); 7.5 months (12.5 menstrual day); 8 months (sixteenth or post-ovulatory day); 8.5 months (twentieth menstrual day); 9 months (twenty-fifth or premenstrual day). Most characteristic is the development of the uterine glands. Starting after the fifth month as small epithelial depressions, they

prenatal life

grow as straight tubes into the soft mesenchyme of the mucosa. Attaining maximal length at eight months, they begin to coil. At the same time the mucosa becomes edematous, and glycogen begins to be deposited. Obviously the luteal factor has become added to the estrogen response. At nine months the endometrium is in a hypertrophic, quasi-premenstrual condition. The blood vessels are heavily congested, and the coiled glands are actively secreting. As already indicated, a precipitous reduction follows the withdrawal of placental hormones at the time of birth.

Endocrine Factors in Sexual Differentiation

If estrogens should play any determining role in the course of normal embryonic differentiation of the female sex organs, apparently no evidence of it has ever been reported. Even considering that the feto-placental unit and the maternal environment make it impossible to raise controls under hormone-free conditions, indications are that tissue responsiveness anyway becomes established only from the fourth month on. This is after the end of embryonic differentiation. The now appearing hormonal effects are of an entirely different nature, belonging to the functional maintenance class.

Quite differently, the embryonic development of male organs is under testicular, androgenic control. The large part of the entire gonaduct system is androgen responsive, probably even preceding the stages of testicular inductive activity. Not only is normal male development dependent on constant hormonal guidance throughout the embryonic period, but female development can progress normally only in the absence of androgens (Jost[9]). Under the influence of deviating endocrine conditions an always persistent androgen-responsiveness may come into evidence expressing itself in the various abnormalities of hermaphordism and adrenogenital syndromes. According to the stage at which interference occurs, various abnormal types will result. Of particular interest are the frequent cases of arrest of the downgrowth of the vagina and a consequently persisting opening into the urethra (Witschi,[17] Wilkins[15]). Possibly the vaginal bud, because of its component of mesonephric duct blastema, is particularly sensitive—as is also the phallic rudiment—to androgen impact. In proportion to the height of attachment at the urethra, teratogenesis must have started earlier or later in fetal life, but always between two and one-half months and four months.

development of the uterus

Discussion

For nearly half a century discussions on the origin of the human vagina have started with enumeration of three or in fact five different views expressed in the literature. Repeatedly suggested as source materials were the oviducts, the mesonephric ducts, and the U.G. sinus epithelium or combinations of some of these. Coverage having been abundant, reference may be made only to the recent papers of Bulmer[2] and Forsberg[6] for bibliographic guidance. The difficulty of reaching a clear decision on the relative value of these proposals stems from an initial misunderstanding.

On the So-called Uterovaginal Rudiment

In the past it had always been assumed that in 25 to 35 mm embryos the lower oviducts, fusing within the fork of the two mesonephric ducts, gave origin to a uterovaginal rudiment. Consequently, whatever elements were added, to line or replace the vaginal segment, supposedly were to be pushed upward from a thin tissue lamella at the lower end of the rudiment. To overcome this time-honored tenet of the uterovaginal rudiment it was necessary only to apply a ruler and to plot the progress of a gonaduct development on graph paper. At once it became evident that the vagina grows not upward but down. It does so in relative distances to landmarks such as the tops of genital cords, orifices of ureters, low rims of the pubic symphyses, and entrances of the vestibular glands. The prostatic buds are indistinct markers in female fetuses, but still help to confirm that the lower end of the vagina is sliding down along the urethra to its separate opening into the vestibulum (Fig. 6).

Koff,[10] avoiding the assumption of contradictory behavior of the mesonephric ducts—namely to proliferate in upward direction after having regressed all the length from above the gonads down to the tubal ends—believed that the mesonephric ducts completely disappear. Simultaneously and in almost exactly the location where the duct orifices had opened, two endodermal pouches (sino-vaginal bulbs) would push out of the dorsal wall of the U.G. sinus. Fusing and penetrating into the "utero-vaginal" rudiment, they then should displace much of the oviducal epithelium and give origin to the lower third of the final vaginal epithelium. Bulmer,[2] though aware that the lower mesonephric ducts and tubal orifices are quite durable structures,

prenatal life

nevertheless supports and further extends the notions of Koff. According to him the "sinus upgrowth" starting from a "differentiated epithelium in the dorsal wall of the sinus" passes through the entire vagina to furnish its final epithelial lining. Fluhmann[5] even favors the idea that also the entire cervical mucosa might be of endodermal (sinus) origin. Davies and Kusama[3] cautiously reserve judgment about the "precise limits" of oviducal and sinus contributions to the cervix. Mysberg[11] comes close to recognizing the important role played by the mesonephric ducts, but also is enmeshed in old misconceptions.

The Vagina Bud

Now that it is entirely clear that the (mesodermal) uterine rudiment never pierces through the wall of the sinus and does not become invaded by endodermal or any other sort of tissue through its lower end, further consideration should be given to the formation of the vagina bud. At the time of its appearance it is composed of a mesial mass of oviduct blastema (the lower, solid tip of the epithelial uterine rudiment) and two lateral pads of mesonephric duct blastema. The mesonephric ducts started to fuse with the tips of the paired oviducts already at stage 33 (25 mm Fig. 3). This divides the ducts into three parts. The ones above the juncture usually disappear early, but may occasionally persist into adult life. Physiologically such remnants seem to be of no importance. The parts below the juncture are short terminal tubules with flaring funnels. As early as at stage 30 (14.6 mm) they have introduced a mesodermal plaque into the dorsal wall of the U.G. sinus. During the time of the formation of the vagina bud the ostial fields from right and left are broadly contiguous and form a spreading epithelium of high cylindric cells. It has attracted the interest of several authors, but only Forsberg[6] seems to have recognized its true mesodermal nature. The persisting short tubes and funnels serve to anchor the bud and to guide the descent of the vagina dorsally behind the urethra. The mechanics of displacement of the funnels down the wall of the urethra is probably similar to that of the ureters toward the bladder at an earlier stage. Finally, the middle sections of the mesonephric duct remnants disaggregate progressively in round or polyhedral blastemic cells, grouped in pads intimately applied to the sides of the central plug of oviduct blastema.

development of the uterus

The formation of such a composite vagina bud might be hard to understand and accept if the entire early history had not already shown that mesonephric duct and oviduct are closely related tissues, both of nephric origin, and the tips of the oviducts always having been attached to the mesonephric ducts during their downgrowth from the upper mesonephroi into the pelvic region (Fig. 1).

Specificity of Tissue Character and Hormonal Responsiveness

Although this presentation concentrates on human materials, it is useful to refer to some observations on the salamander Ambystoma.[17] In this species, as in man, the oviduct grows and separates from a tip imbedded in the lateral wall of the mesonephric duct. By testosterone stimulation one may stop further downgrowth of the oviduct at any chosen level. This reaction obviously is analogous to that of the above-mentioned virilization in human fetuses. Of particular interest in the salamander experiment is an immediate change in hormone responsiveness that occurs in the mesonephric duct as soon as the oviduct has separated off. It is not unlikely that in the human second fetal period the major reaction difference to estrogen stimulation— keratinization of the vaginal versus mucification of the uterine epithelia—is bound up with the fact that only in the formation of the vagina, but not in the uterus, mesonephric duct and oviduct elements remain united. We should not, however, stretch this possibility of explanation too far, since tissue specificity has no inflexible limits. Borders of cornification indeed may shift. I am inclined to agree with Bern[1] when he says "the ability to keratinize under some circumstances may be a fundamental property of the basal cells of all epithelia in mammals." Davies and Kusama[3] refer to the chemical similarity of keratin and mucus which both are sulfur-containing proteins. Rosa[12] and Fosberg[6] have gone far in the analysis of shifting chemical and enzymatic conditions which accompany morphologic changes during sexual differentiation. They furnish essential materials for the scientific understanding of embryologic mechanics. Histochemistry has always been valuable in the study of developmental processes. It will have to lead to the understanding of the nature and the origin of tissue and organ specificity. However, extreme caution should be exerted in its application to problems of cell migration.[19]

prenatal life

Acknowledgments

The author is grateful to Dr. James D. Ebert, Director of the Department of Embryology, Carnegie Institution, for his welcome at the new Baltimore laboratories and permission to use the Human Embryo Collection for study and publication. Thanks are extended also to Dr. Bent G. Böving and photographer R. D. Grill for kind interest and valuable assistance during my visit.

References

1. Bern, H. A.: Epithelial metaplasia in the prostate and other genital structures of male mammals. *Nat Cancer Inst Monogr, 12:* 43-45, 1962.
2. Bulmer, D.: The development of the human vagina. *J Anat, 91:* 490-508, 1957.
3. Davies, J., and Kusama, H.: Developmental aspects of the human cervix. *Ann NY Acad Sci, 97:* 534-550, 1964.
4. Felix, W.: The development of the urogenital organs. In F. Keibel and F. P. Mall, *Manual of Human Embryology,* Philadelphia, Lippincott, 1912, vol. II, pp. 752-979.
5. Fluhmann, C. F.: The developmental anatomy of the cervix uteri. *Obstet Gynec, 15:* 62-69, 1960.
6. Forsberg, J. G.: Origin of vaginal epithelium. *Obstet Gynec, 25:* 787-791, 1965.
7. Halban, J.: Schwangerschaftsreationen der foetalen Organe und ihre puerperale Involution. *Z Geburtsh Gynaek, 53:* 191-231, 1904.
8. Hunter, R. H.: Observations on the development of the human female genital tract. *Carnegie Contrib Embryol, 22:* 91-107, 1930.
9. Jost, A.: Recherches sur le contrôle hormonal de l'organogenèse sexuelle du lapin et remarques sur certaines malformations de l'appareil génital humain. *Gynec Obstet* (Paris), *49:* 44-60, 1950.
10. Koff, A. K.: Development of the vagina in the human fetus. *Carnegie Contrib Embryol, 24:* 59-90, 1933.
11. Mijsberg, W. A.: Ueber die Entwicklung der Vagina, des Hymen und des Sinus urogenitalis beim Menschen. *Z Anat Entwicklungsgesch, 74:* 684-760, 1924.
12. Rosa, P.: *Endocrinologie sexuelle du foetus féminin.* Paris, Masson, 1955.
13. Scammon, R. E.: The prenatal growth of the human uterus. *Proc Soc Exp Biol Med, 23:* 687-690, 1926.
14. Schwers, J.: *Les oestrogènes au cours de la second moitié de la grossesse.* Hôpital Universitaire St. Pierre. Bruxelles, Editions Arscia S.A., 1964.
15. Wilkins, L.: *The Diagnosis and Treatment of Endocrine Disorders in Childhood and Adolescence.* 2d ed., Springfield, Thomas, 1957.
16. Witschi, E.: *Development of Vertebrates.* Philadelphia and London, Saunders, 1956, pp. 526-51.

17. ———: Embryology of the uterus: –normal and experimental. *Ann NY Acad Sci, 75*: 412-435, 1959.
18. ———: "Grundlagen der sexuellen Differenzierung." In Käser, O. *et al.*: *Gynäkologie und Geburtshilfe, 1*: 51-70, Thieme, Stuttgart, 1969.
19. Zuckerman, S.: The histogenesis of tissues sensitive to oestrogens. *Biol Rev, 15*: 231-272, 1940.

chapter II

Placental Circulation in Rhesus and Man

Elizabeth M. Ramsey, M.D., Sc.D.

We shall examine a motion picture which demonstrates, radiologically, the maternal and fetal circulation of the placenta of the rhesus monkey as we have observed it.

The question which immediately comes to the mind of clinicians such as yourselves is: Does information derived from reproductive studies in monkeys have any practical value for us? This question goes to the heart of an important problem. How are we to study the anatomy and physiology of human reproduction when it is inappropriate to apply to human patients some of the necessary experimental procedures? The use of rabbits, mice, and guinea pigs, from which we may obtain much basic biological information, is not the full answer, for these animals are profoundly different from human beings in many pertinent aspects. The monkey, as a primate with a hemochorial placenta, is obviously closer to the human, but how much closer? Is it close enough?

It is forty years since George Corner, Sr. established the first American breeding colony of rhesus monkeys in the Johns Hopkins Department of anatomy.[1] That colony, as continued by Carl Hartman at Carnegie, was the ancestor of the one we use today. In the forty years this same question has been posed and answered in the affirmative for one aspect of reproductive physiology after another. But it is important to note, and we may do so with satisfaction, that the question has been asked anew as each fresh field of

prenatal life

investigation was entered. Carl Hartman asked it with respect to menstruation;[2] Corner, Bartelmez, and Hartman asked it with respect to ovulation, corpus luteum formation, and anovulatory cycles;[3] Bartelmez asked it about the form and the functions of the endometrial spiral arteries in the menstrual cycle;[4] Streeter and Heuser, about early embryonic development;[5] Wislocki and Streeter, about implantation and placentation.[6] And we in our turn ask it again today. What about placental circulation? Is the rhesus monkey a dependable experimental model for that? In point of fact, my colleagues and I posed the question a good many years ago, and it is only because we have been investigating the matter intensively in the interval that I feel justified in suggesting a reply.[7]

Our answer is again an affirmative one. But let us be clear exactly what the question is. We ask: Is rhesus a reliable experimental model? Now, a model is by definition not the original itself nor identical with it, and between our monkey model and its human original there are indeed species differences. We have devoted much time and thought to the identification and evaluation of these differences, and I should like to consider some of the important ones so that you will understand the basis upon which we have reached our affirmative conclusion.

The first difference becomes apparent as early as the time of implantation. The monkey blastocyst attaches upon the surface of the endometrium, and the human blastocyst penetrates beneath it to become implanted within the endometrium (Fig. 1). In addition, a secondary placenta forms in the monkey, a little after the primary site is established, at the point where the enlarging blastocyst comes in contact with the opposite uterine wall.

The next difference concerns decidua formation which is so prominent a development in human gestation but is absent in the monkey (Figs. 1 and 2). The role of nutrition for the implanting ovum which has been ascribed to the human decidua may perhaps be filled by the epithelial plaque in the monkey (Fig. 1A). This is a transitory proliferation of the surface epithelium adjacent to the two placental areas that reaches its peak around the fourteenth day postconception and wanes thereafter. There is, of course, no analogous formation in the human (Fig. 1B).

The behavior of the trophoblast constitutes another difference between the human and the monkey. In man this tissue penetrates rapidly and deeply as evidenced by the irregularity of the fetal-maternal junction

placental circulation

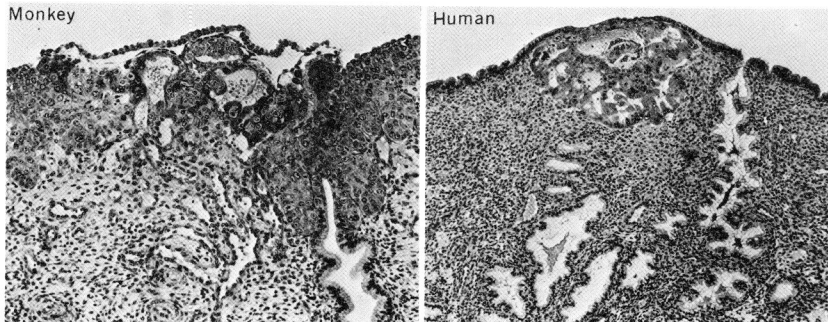

Figure II-1. Sections through monkey and human blastocysts shortly after attachment to the endometrium. The superficial implantation of the monkey contrasts with the interstitial implantation in man. The monkey's epithelial plaque and the early human decidual reaction are apparent. Trophoblastic lacunae connecting with maternal capillaries may be seen in both. Monkey C-524, tenth day after ovulation; section 3-3-10. ×105. Human specimen 8004. Eighth day after ovulation; section 11-4-4. ×85. (From Ramsey and Harris.[7])

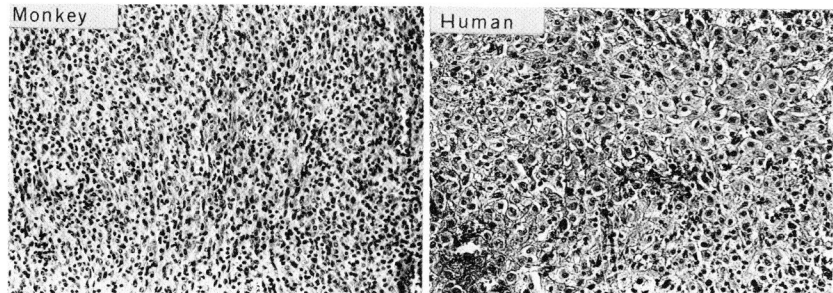

Figure II-2. Sections through the subplacental endometrium in monkey and man to show absence of decidual reaction in the former. Monkey C-658, thirty-eighth day of pregnancy; section S1. ×100. Human specimen 9742, second month of pregnancy. ×100.

prenatal life

(Fig. 3), the wide dissemination of trophoblastic wandering cells, and the replacement of much of the wall of the uteroplacental arteries by trophoblast (Fig. 4). Only the latter is duplicated in rhesus, and it is significant that the trophoblastic cells travel only via the vascular lumina in the monkey, whereas in man they apparently reach the vessel wall both from within and from without. Furthermore, the replacement of the wall is less extensive in rhesus.

Whether the difference in depth of implantation is related to the foregoing is a field of active current investigation. Some authors consider the epithelial plaque to be a barrier to penetration and therefore a causal factor in the monkey's superficial implantation. Others ascribe the difference to the marked variation in trophoblastic invasive properties as between the two species. We cannot here enter into this interesting debate, but only note it as we pass to our major matter of concern: What do all these factors have to do with placental circulation? Do they occasion significant circulatory differences between monkey and man?

A morphologic study of the transformation of uteroplacental vessels throughout pregnancy helps us to answer this question. If representative arteries from the two species are modeled at comparable stages of gestation the overwhelming similarity in the changes they undergo becomes obvious (Fig. 5). There is less stretching out of coils in human arteries at term and a greater accentuation of the dilated, terminal sac just at the point of entry to the intervillous space. The former is perhaps related to the greater thickness of the human uterine wall and the latter to the more extensive replacement of the arterial wall by trophoblast. But from the standpoint of circulatory mechanism, we cannot say that these qualitative variations upon the consistent basic theme are significant. Morphologically our "experimental model" seems to be valid.

In the realm of physiology there is another parameter to be considered, namely, the effect upon placental circulation of myometrial contractions. Fundamental to investigation of that factor is the question whether the monkey uterus has the same pattern of activity as the human.

Again we place side by side representative monkey and human data. The monkey recordings come from our own studies, the human ones were kindly given us by Dr. Charles Hendricks (Fig. 6). The work of Hendricks,[8] Caldeyro-Barcia,[9] and others has firmly established the pattern of activity in human pregnancies. It can be seen that the only important variation lies

placental circulation

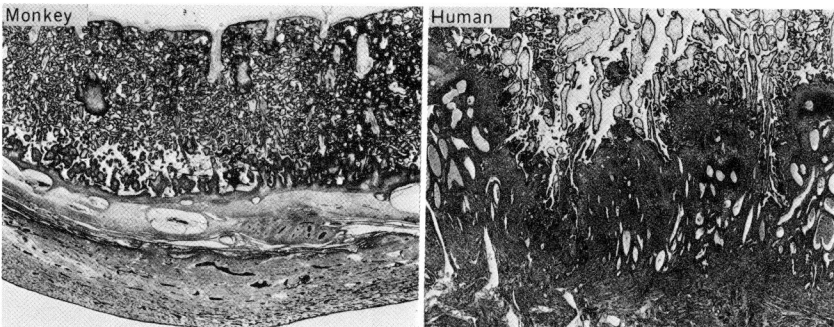

Figure II-3. Sections to show the difference in maternal-fetal junction; smooth in the monkey, strikingly irregular in man. In some areas in the human specimen the trophoblast penetrates nearly to the myometrium. Traces of injected India ink can be seen in the maternal vessels and the intervillous space in the monkey specimen. Monkey C-629, fifty-third day of pregnancy; section 32. ×8. Human specimen 10129, sixty-first day of pregnancy; section 2. ×8. (From Ramsey and Harris.[7])

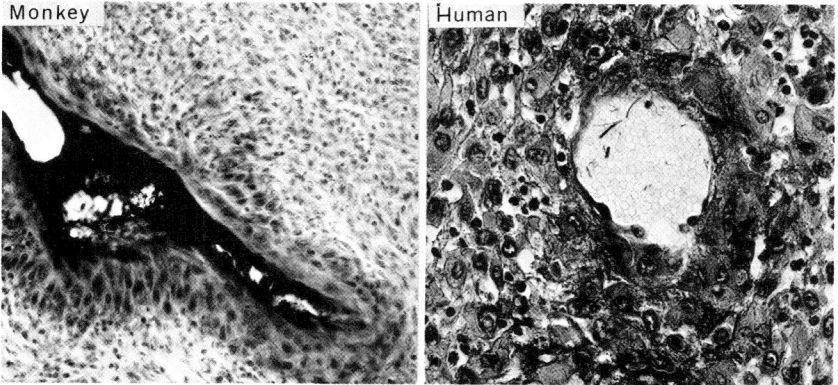

Figure II-4. Uteroplacental arteries near the placental base showing trophoblastic penetration of the vessel wall. In the monkey there is no trophoblast in the lumen of the artery or in the endometrial stroma. Trophoblast appears in both these locations in man. Injected India ink is present in the lumen of the monkey artery. Monkey C-629, fifty-third day of pregnancy; section 100. ×200. Human specimen 10117, eighty-fifth day of pregnancy. ×400.

prenatal life

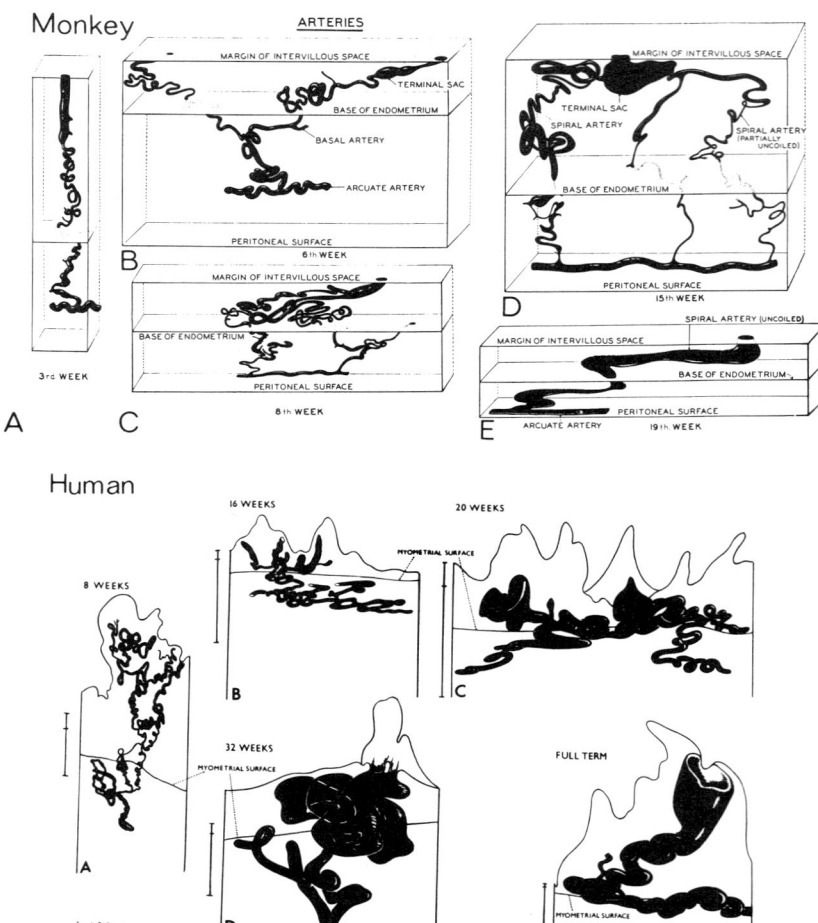

Figure II-5. Diagrammatic representations of the course and configuration of the uteroplacental arteries in monkey and human gestation, based on three-dimensional models constructed from serial sections. Comparable stages of pregnancy are shown. Ratio of monkey gestation period to human, approximately 5:8. (From Ramsey and Harris.[7])

placental circulation

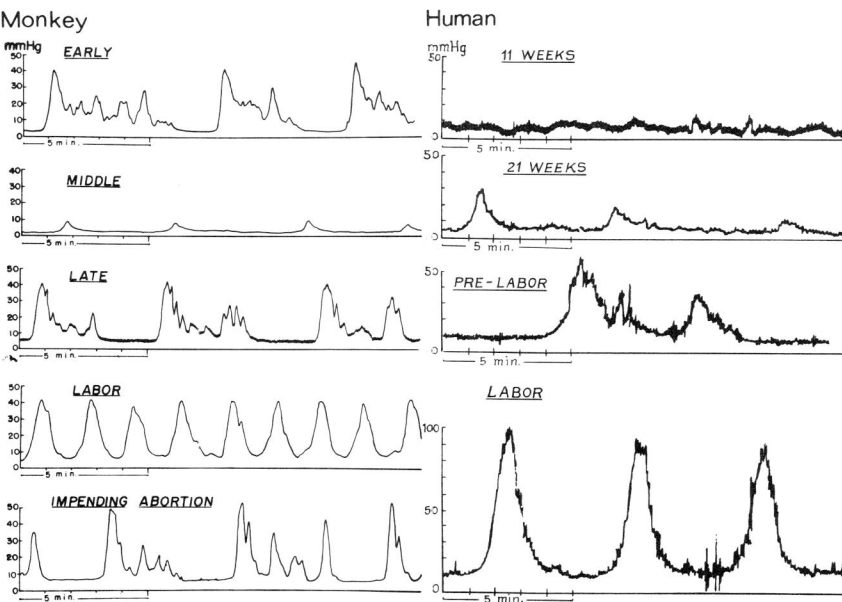

Figure II-6. Representative recordings of intrauterine pressure in pregnant monkeys and women at three stages of pregnancy and in labor (human recording, courtesy of Dr. Charles H. Hendricks). (From Ramsey and Harris.[7])

in the early stage of pregnancy when the monkey uterus is more active than the human. This may be accounted for by one or more strictly monkey conditions such as lesser production of placental progesterone, the bidiscoid placenta, or simply the disproportion between total uterine size and myometrial thickness at this early stage. In any case, the important feature from the standpoint of circulation is the occurrence in both species of uterine contractions of similar overall pattern. The contractions are characterized by increasing intensity, frequency, and coordination as term approaches and labor progresses.

The final demonstration of human-monkey similarity is provided by radioangiography of placental circulation. Borell[10] and his associates in

prenatal life

Sweden have been the leaders in radioangiography of the human placenta. Our observations in the monkey are illustrated by the motion picture I shall show you. First, however, let me present one more example of side-by-side data for monkeys and humans (Fig. 7). In these two still radiograms the same pattern of filling of uteroplacental arteries, following femoral arterial injection of the contrast medium, may be seen in both species. The endometrial spiral arteries are prominent and from their distal ends the contrast medium is entering the intervillous space in characteristic fountainlike spurts. I think we would all agree that the two subjects can only be differentiated on the basis of size.

This brief analysis and evaluation of species differences as related to placental circulation shows that it is reasonable to look upon our experimental model as dependable and, in viewing the film, to regard it as representative of the primate type (Fig. 14).

Summary of Script of Motion Picture "Placental Circulation in the Rhesus Monkey," E. M. Ramsey, M.D., M. W. Donner, M.D., and C. B. Martin, Jr., M.D. With the advice and technical assistance of B. G. Böving, M.D., H. R. Misenhimer, M.D., S. I. Margulies, M.D., and R. D. Grill, B.P.A.

In the radiographic studies two modalities were utilized.[11] In some studies a system was employed which provides standard X-rays made in two planes simultaneously at a rapid rate. Our studies with this equipment were conducted at a speed of two films per second. In other studies we used cineradiography in which a motion picture camera photographs the intensified fluoroscopic image at a speed of thirty frames per second.

The rapid serial X-rays provide excellent definition, and the cineradiography permits observation of circulatory dynamics. Thus the two techniques are complementary.

To demonstrate maternal arterial inflow to the placenta a standard femoral cut down is performed and a catheter is introduced into the femoral artery to convey the radiopaque material into the maternal circulation. The contrast material is injected as bolus with a pressure syringe.[12]

The contrast material progresses from the femoral artery into the aorta, rising to the level of the renal arteries. It then returns under maternal

placental circulation

arterial pressure to the uterine arteries (Fig. 8). The endometrial spiral arteries are filled and the medium enters the intervillous space in fountainlike spurts. These spurts grow and eventually coalesce as the contrast material spreads in the intervillous space of the placenta.

To demonstrate venous drainage from the placenta, contrast material is injected into the intervillous space.[13] In approximately 80 percent of

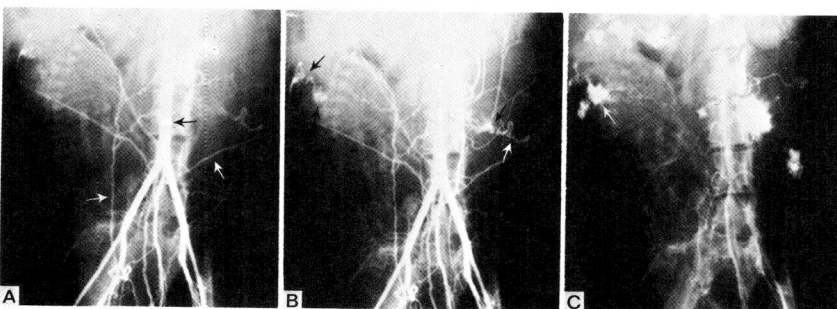

Figure II-8. Angiograms made one, two, and three seconds, respectively, after injection of contrast medium into the femoral artery. In *a*, the upper arrow points to the aorta, the lower ones to the uterine arteries. In *b*, the two outer arrows indicate the spiral endometrial arteries, the two inner arrows, "spurts" into the intervillous space. The arrow in *c* also shows a "spurt." Monkey 60/14. SP#3, 100th day of pregnancy.

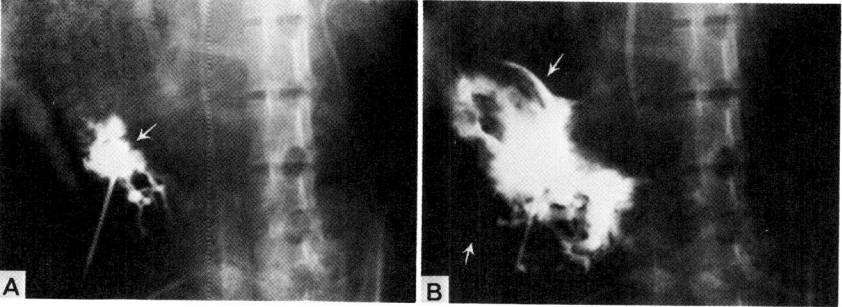

Figure II-9. Angiograms made at two and six seconds after rapid injection of contrast material into the intervillous space during uterine relaxation. The arrow in *a* points to the pool of opaque medium in the intervillous space. The two arrows in *b* indicate mural veins. Monkey 62/26. SP#14, 124th day of pregnancy.

prenatal life

rhesus monkeys the placenta is bidiscoid. The location of the placental discs is ascertained by transillumination of the uterus, and a catheter is introduced into the intervillous space of one of the discs. Through this the radiopaque medium is injected with a constant speed infusion pump.

The contrast material slowly disperses throughout the intervillous space, and drainage channels in the uterine wall appear in increasing numbers as the space fills (Fig. 9).

The preceding illustrations have shown uteroplacental circulation during uterine relaxation. The effects of contractions upon circulation will now be considered.

In order to inject the contrast medium at the desired time, it is necessary to monitor uterine activity continuously. This is accomplished by the introduction of an open-end catheter into the amniotic cavity by transabdominal amniocentesis. The catheter is attached to a pressure transducer and the latter to a multichannel physiological data recording system.[14] The X-ray procedures are similar to those already described.

During uterine contraction the arterial inflow is reduced, or even halted, depending upon the intensity of the contraction.[12] In the illustration in Figure 10, showing an experiment in which the medium was injected during an intense contraction, there are no spurts and there is only partial filling of the uterine vasculature.

Uterine activity also affects venous drainage of the intervillous space (Fig. 11).[13] As the uterus begins to contract the venous drainage channels disappear or become narrowed. However, the pool of contrast material within the intervillous space remains essentially the same throughout the contraction. This suggests that contraction does not empty the intervillous space. As the uterus begins to relax, venous channels reappear.

Thus far only half of the placental circulation has been considered. No mention has been made of the contribution of the fetus.

Demonstration of the fetal circulation in the placenta of the monkey can be accomplished in one of two ways, either by injection of contrast material into the fetal femoral artery[15] or by direct injection into an interplacental vessel.[16] Both of these routes have been employed.

To inject radiopaque material into the femoral artery of the fetus laparotomy is performed. The uterus is transilluminated to outline the margins of the placental discs so that these may be avoided. The uterine

placental circulation

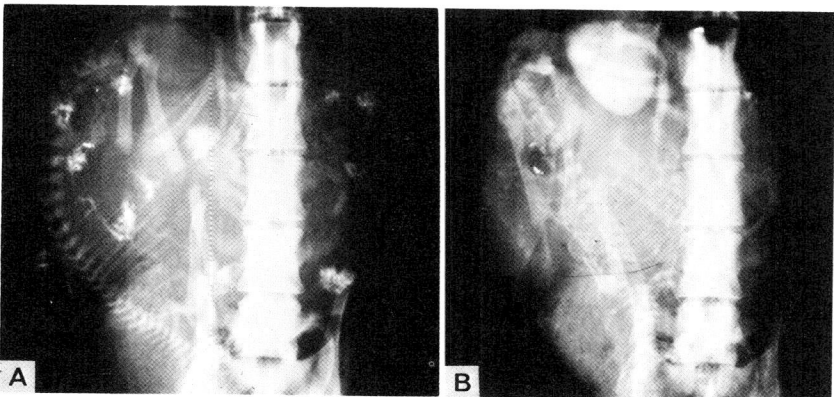

Figure II-10. Angiograms made eight seconds after injection of contrast medium into the femoral artery. a) Injection during uterine relaxation; b) injection during a contraction of 42 mm. Monkey 62/30. SP#38, 141st day of pregnancy.

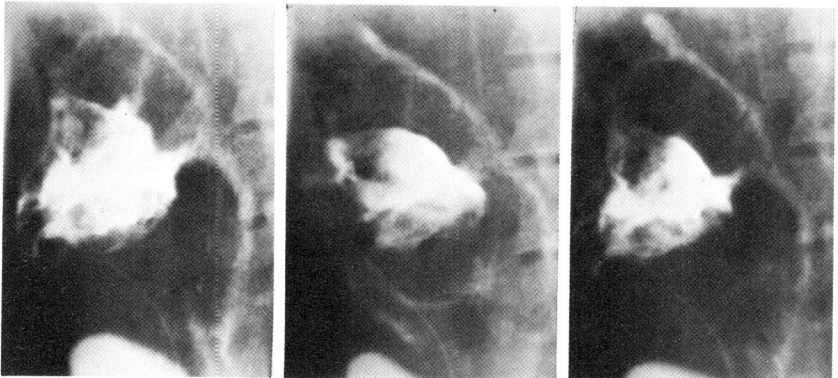

Figure II-11. Photographs of three areas from a cineradiogram made immediately after the conclusion of a slow infusion of contrast material into the intervillous space. The right and left radiograms were made during relaxation, the center radiogram during the contraction which occurred between the two relaxation periods. Some drainage channels disappear during the contraction and others are narrowed. Refilling of channels occurs with return of relaxation. Note maintenance of size of pool in intervillous space. Monkey 62/29. MP#39. Ninety-ninth day of pregnancy.

prenatal life

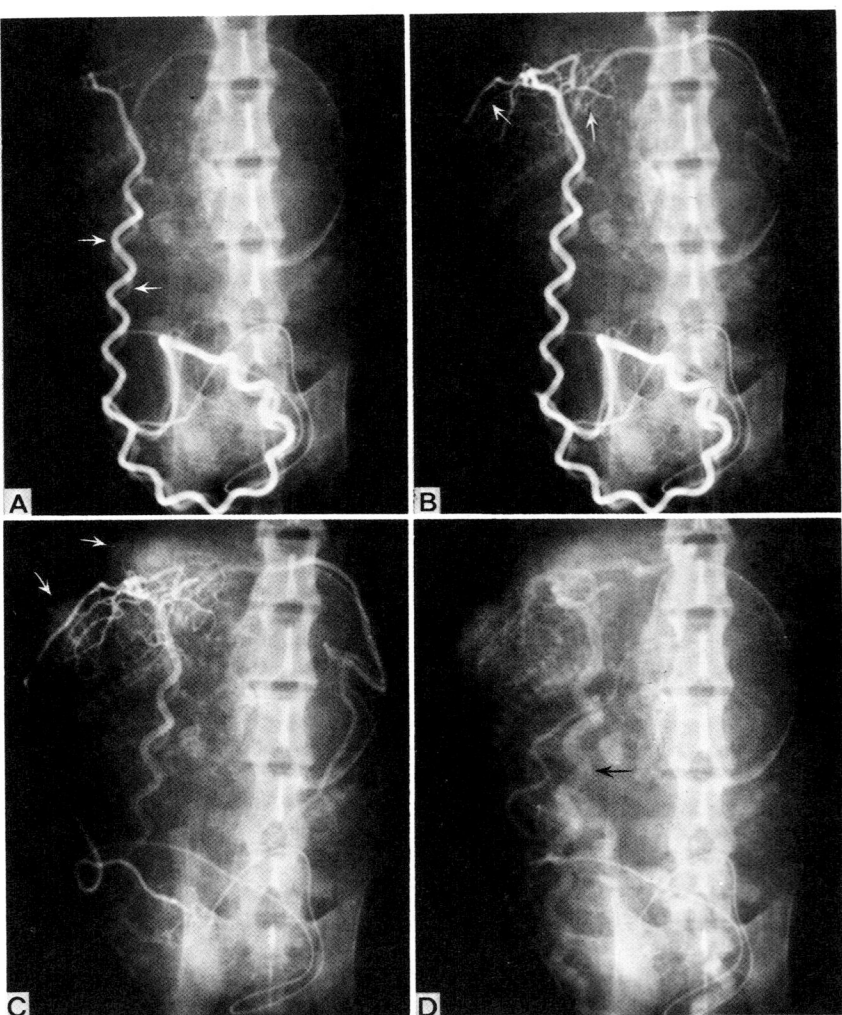

Figure II-12. Four of a series of rapid serial radiograms made two, two and one-half, four, and nine seconds, respectively, after injection of contrast material into the fetal femoral artery. The arrows in *a* point to the two umbilical arteries; those in *b* to the subchorionic vessels; those in *c* to two of the fetal cotyledons. By the time radiogram *d* was made contrast material had entered the umbilical vein (dark arrow). Monkey 64/72. SP#22, 138th day of pregnancy.

placental circulation

wall is then incised. The membranes are separated from the wall in the region of the incision. A fetal leg is manipulated through this opening, with the membranes intact when possible. The fetal groin is exposed, and an incision is made over the femoral vessels. These are identified, and the femoral artery is isolated and incised to permit the introduction of a small polyethylene catheter. When this has been accomplished, there is prompt return of blood under the head of fetal arterial pressure.

Radiopaque material injected into the fetal femoral artery returns to the fetal hypogastric arteries and the umbilical arteries (Fig. 12). It rises to the level of the chorionic plate and diffuses into the subchorionic vessels.

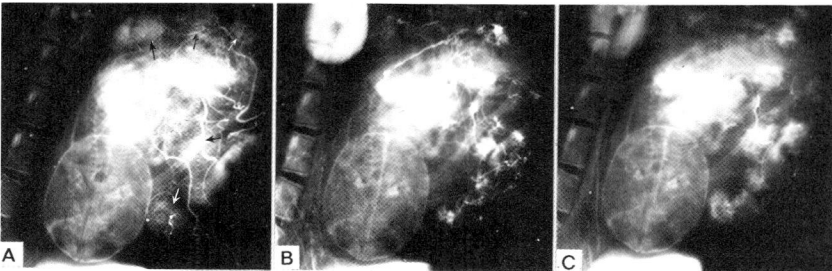

Figure II-13. Spot films made during a combined fetal and maternal injection study. a) Three seconds after injection of contrast material into an interplacental artery (arrow at center right). The fetal cotyledons appear as cottony puffs (upper and lower arrows). b) Two seconds after an immediately subsequent maternal injection. "Spurts" of contrast material from maternal arteries overlie the cotyledons. In c, made six seconds after b, the "spurts" are enlarging and becoming confluent. Monkey 65/80. MP#59, 152nd day of pregnancy.

Then it slowly fills the capillaries of the fetal cotyledons which appear as discrete, cottony puffs. The cotyledons are independent structures. There is no capillary communication from one to the other. From the cotyledons the blood drains into the umbilical vein and is returned to the fetal circulation.

As previously mentioned, the placenta of the rhesus monkey is usually bidiscoid, and the two discs are connected by fetal vessels. These interplacental vessels can be located by transillumination of the uterus. Through a hysterotomy incision they are dissected free and catheterized to permit in-

prenatal life

jection of contrast material directly into the subchorionic circulation. This technique provides superior visualization of the fetal cotyledons and of the placental disc which has been injected.

Contrast material enters the interplacental vessel and appears in the capillary network of the fetal cotyledons (Fig. 13).

There have been presented, in sequence, the methods used to study the two components of placental circulation, the maternal and the fetal. The final demonstration shows these two independent circulations simultaneously to illustrate their relationship.

The radiopaque material is first injected into the fetal circulation and, immediately thereafter, a maternal injection is begun.[16] In Figure 13 the X-ray on the left was made a few seconds after a fetal injection. The radiopaque material appears in the fetal capillary system of the cotyledon.

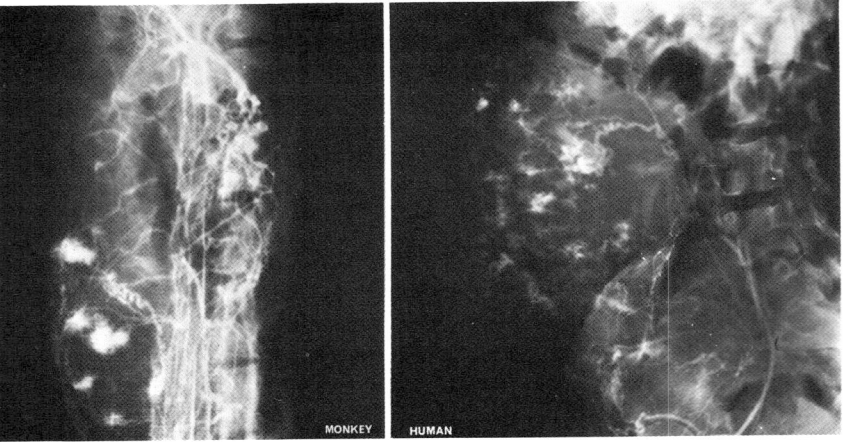

Figure II-7. Radiographs made following injection of contrast material into the femoral artery. In both monkey and man the opaque medium is seen in the myometrial arteries and in the endometrial spiral arteries. "Spurts" of opaque medium at the end of the arteries in both subjects mark the entry of the medium into the intervillous space. Monkey 1B, 117th day of pregnancy. Human specimen, University of Virginia No. 372656; thirty-second week of pregnancy (courtesy of Dr. Harry S. McGaughey, Jr.) (From Ramsey and Harris.[7])

placental circulation

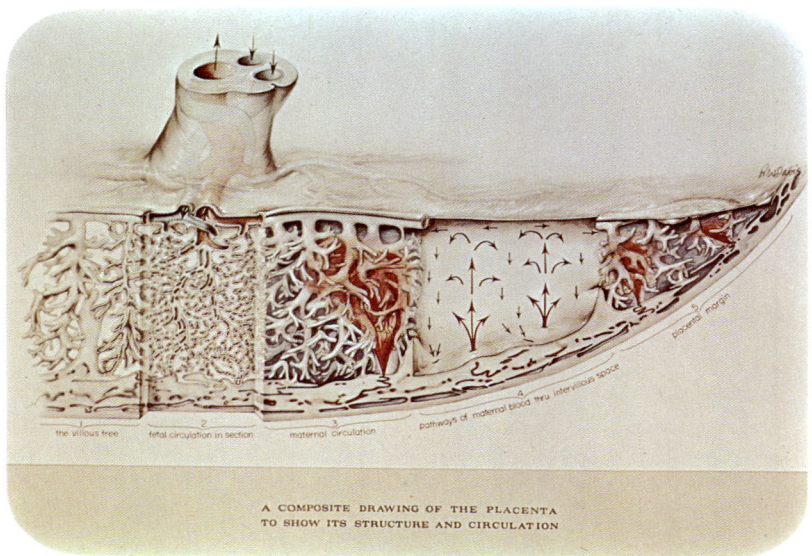

Figure II-14. A composite drawing of the placenta to show its structure and circulation. Drawing by Ranice Davis Crosby (courtesy of the Carnegie Institution of Washington).

The film in the center shows an early stage of an immediately subsequent maternal injection. Spurts into the intervillous space are beginning to be visible. The final X-ray shows well formed spurts from the maternal circulation directly overlying fetal cotyledons in the ratio of 1:1.

Summary

From the motion picture it is apparent that the technique of radioangiography is a suitable one for direct visualization of placental circulation. By employing it we have been able to demonstrate the pathways of inflow of maternal blood to the intervillous space and of drainage of blood back into the maternal systemic circulation.

prenatal life

We have shown that myometrial contractions curtail or halt both arterial inflow and venous drainage.

Finally we have demonstrated the basic relationship between maternal and fetal circulation in the placenta of the rhesus monkey.

References

1. Corner, G. W.: Ovulation and menstruation in *macacus rhesus*. *Carnegie Contrib Embryol, 15:* 73-101, 1923.
2. Hartman, C. G.: Studies in the reproduction of the monkey *macacus (pithecus) rhesus*, with special reference to menstruation and pregnancy. *Carnegie Contrib Embryol, 23:* 1-161, 1932.
3. Corner, G. W., with the collaboration of C. G. Hartman and G. W. Bartelmez: Development, organization, and breakdown of the corpus luteum in the rhesus monkey. *Carnegie Contrib Embryol, 31:* 117-146, 1946.
4. Bartelmez, G. W.: The form and the functions of the uterine blood vessels in the rhesus monkey. *Carnegie Contrib Embryol, 36:* 153-182, 1957.
5. Heuser, C. H., and Streeter, G. L.: Development of the macaque embryo. *Carnegie Contrib Embryol, 29:* 15-55, 1941.
6. Wislocki, G. B., and Streeter, G. L.: On the placentation of the macaque (*macaca mulatta*), from the time of implantation until the formation of the definitive placenta. *Carnegie Contrib Embryol, 27:* 1-66, 1938.
7. Ramsey, E. M., and Harris, J. W. S.: The morphology of human uteroplacental vasculature. *Carnegie Contrib Embryol, 38:* 59-70, 1966.
8. Hendricks, C. H.: The hemodynamics of a uterine contraction. *Amer J Obstet Gynec, 76:* 969-982, 1958.
9. Alvarez, H., and Caldeyro-Barcia, R.: Fisiopatologica de la contraccion uterina y sus aplicaciones en la clinica obstetrica. Presented at Segundo Congreso Latinamericano de Obstetrica y Ginecologia and at Cuarto Congreso Brasilero de Obstetrica y Ginecologia, San Pablo, 1954.
10. Borell, U., Fernström, I., Ohlson, L., and Wiqvist, N.: Effect of uterine contractions on the human uteroplacental blood circulation: An arteriographic study. *Amer J Obstet Gynec 89:* 881-890, 1964; Influence of uterine contractions on the uteroplacental blood flow at term. *Amer J Obstet Gynec, 90:* 44-57, 1965a.
11. Donner, M. W., Ramsey, E. M., and Corner, G. W., Jr.: Maternal circulation in the placenta of the rhesus monkey: A radiographic study. *Amer J Roentgen, 90:* 638-649, 1963.
12. Ramsey, E. M., Corner, G. W., Jr., and Donner, M. W.: Serial and cineradioangiographic visualization of maternal circulation in the primate (hemochorial) placenta. *Amer J Obstet Gynec, 86:* 213-225, 1963.

13. Ramsey, E. M., Martin, C. B., Jr., McGaughey, H. S., Jr., Kaiser, I. H., and Donner, M. W.: Venous drainage of the placenta in rhesus monkeys: Radiographic studies. *Amer J Obstet Gynec 95:* 948-955, 1966.
14. Corner, G. W., Jr., Ramsey, E. M., and Stran, H. M.: Patterns of myometrical activity in the rhesus monkey in pregnancy. *Amer J Obstet Gynec, 85:* 179-185, 1963.
15. Martin, C. B., Jr., Ramsey, E. M., and Donner, M. W.: The fetal placental circulation in rhesus monkeys demonstrated by radioangiography. *Amer J Obstet Gynec 95:* 943-947, 1966.
16. Ramsey, E. M., Martin, C. B., Jr., and Donner, M. W.: Fetal and maternal placental circulations. *Amer J Obstet Gynec, 98:* 419-423, 1967.

chapter III

The Physiologic Response to Uterine Contractions

Charles H. Hendricks, M.D.

For thousands of years mankind has been aware of woman's symptomatic response to the pain of the uterine contractions of labor. More recently we have begun to concern ourselves with the effect of uterine contractions upon the fetus. Seldom, other than in the context of pain, have we stopped to consider the effects of the contraction itself upon the mother. In this presentation we will consider some of the physiological responses made by the woman in labor to her uterine contractions and the implications of those responses.

What Happens When the Uterus Contracts?

Let us assume for the moment that no additional mechanical phenomena are involved and review the current concept of what happens to the circulation when the uterus contracts. At the onset of the contraction it appears that blood is squeezed out of uterine venous sinuses very rapidly,[1,2] approximately 250 to 300 cc being extruded back into the central circulation.[3] The venous outflow appears to be arrested quite early in the contraction cycle through constriction of the veins.[2,4] As the pressure increases, the arterial input undergoes progressive attrition until finally it is shut off as well.[5-7] Thus

Supported in part by USPHS Research Grant HD-00264-13 from the National Institute of Child Health and Human Development.

prenatal life

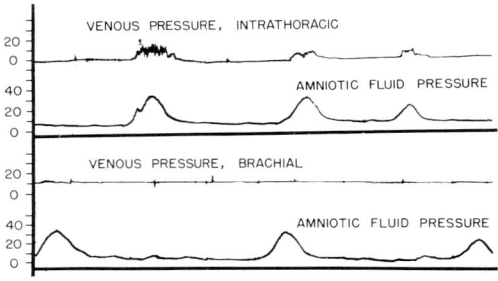

Figure III-1. During a contraction of the intrathoracic venous pressure rises. The brachial venous pressure is unaffected.

at the apex of an effective contraction in labor, blood is neither entering nor leaving the uterus. As the uterus relaxes, the arterial input resumes, but apparently at a somewhat lower level of uterine pressure than the point

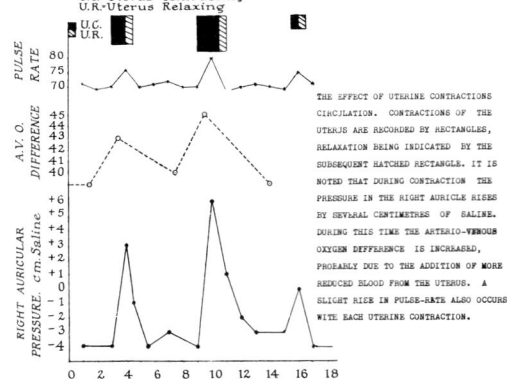

Figure III-2. During the active phase of contractions the heart rate rises, the arteriovenous oxygen difference increases, and the right auricular pressure increases.[1]

at which it had been shut off, for reasons that are not clear. Finally, venous outflow resumes until the contraction cycle is repeated.

The Problem of Redistribution

When the estimated 250 to 300 cc of blood are squeezed rapidly back into the central venous reservoir, it must be dealt with by the central circulation. Basically it appears to be a matter of redistributing this amount of blood into a circulation from which uterine participation has been temporarily excluded. The presence of such a temporary surfeit of blood in the central circulation sets of a logical chain of events.

At first there may be noted a small increase in intrathoracic pressure but not in the brachial pressure (Fig. 1). The increase in central venous pressure is observed regardless of whether or not the contraction is painful. As observed by Palmer and Walker there is also a small but significant increase in the right auricular pressure and also the arteriovenous oxygen difference is significantly increased (Fig. 2).[1]

An initial increase in heart rate followed by a decrease as the contraction advances has been noted (Fig. 3).[3,8] Stroke volume (estimated by the pulse pressure method)[9] undergoes a transient reduction, followed by a substantial increase which reaches its maximum near the apex of the contraction.

The cardiac output (again estimated by the pulse pressure technique) begins to rise within a few seconds after the onset of the contraction and continues a modest rise which usually reaches its peak somewhat before the peak intrauterine pressure is achieved.[3] Figure 4 shows the maximum cardiac output in a series of twenty individual contractions. In Figure 5 we have combined the output responses of all twenty contractions to give a better visualization of the typical output alteration.

Change in blood pressure response in uterine contractions is variable in the normotensive patient. Almost always, however, there is some increase in the systolic and diastolic blood pressure in response to contractions, the former usually increasing more than the latter.[5] Figure 6 illustrates most of the important phenomena of the blood pressure response. This tracing shows a fifty-minute segment of uterine activity of a woman in whom an oxytocin-induced labor was being conducted. In the left half of the illustration the patient was lying on her side. Here there may be seen great variations in the blood pressure rises, the degree of alteration of the blood pressure usually being somewhat in proportion to the size of the uterine contraction, with systolic rises usually within the range of 10 to 20 mm Hg., while those

prenatal life

Figure III-3. With a contraction the heart rate first rises, and then progressively slows as the peak of the intrauterine pressure approaches, returning to its baseline level only after the completion of the contraction. The stroke diminishes slightly, and then increases in response to the contraction, returning to the base level as the contraction is completed.[3]

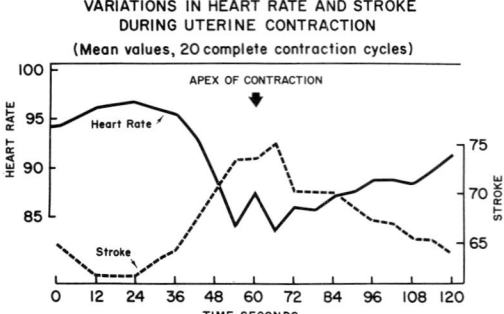

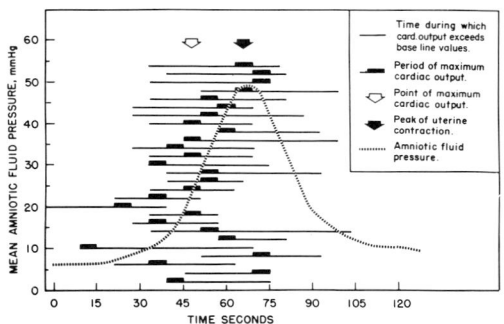

Figure III-4. A composite picture showing the increase in cardiac output (estimated by pulse pressure method) during twenty separate contraction cycles. The heavy oblong black marker represents the period of maximum cardiac output for each of these contractions.[3]

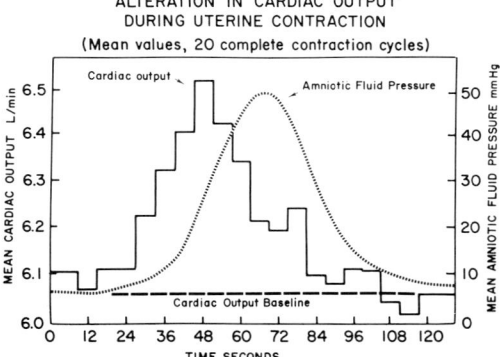

Figure III-5. The mean cardiac output change of the twenty cycles illustrated in Figure 4 is plotted against the time of a uterine contraction. Note that the peak output is reached somewhat before the peak intrauterine pressure.[3]

uterine contractions

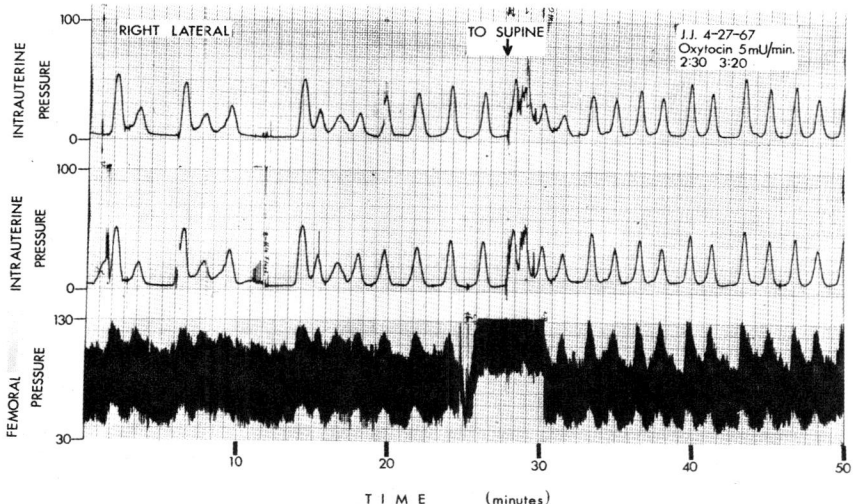

Figure III-6. A tracing of intrauterine pressure and femoral arterial pressure during a fifty-minute period of induced labor. The blood pressure response to uterine contractions is much more marked when the patient is in the supine position.

of the diastolic pressure usually were somewhat less. Midway through this part of the record the patient turned to the supine position after which the blood pressure responses became much more marked and much more consistent, the rise still, however, being roughly proportional to the size of the uterine contraction.

The oscillations in blood pressure illustrated here are greater than those seen in the average patient. Ordinarily there is observed only a modest rise in arterial blood pressure when the patient is in the supine position, and a significantly smaller one when the patient is in the lateral position. If one examines continuous arterial blood pressure tracings carefully, however, one can almost invariably identify blood pressure alterations of some degree in response to each uterine contraction.

At one time it had been suggested that the increase in mean arterial pressure that accompanies the uterine contraction would be of potential value in helping to continue perfusion of the uterus during a uterine con-

prenatal life

traction cycle. This does not appear to be the case, however, since the maximum blood pressure rise tends to correspond fairly closely in time with the peak of the uterine contraction, when the uterus is denying both ingress and egress to the systemic circulation.

It has been known for many years that the femoral venous pressure is elevated when the patient is in the supine position. More recently it has been demonstrated that under these circumstances the femoral venous pressure sometimes rises and falls in almost perfect imitation of a contraction

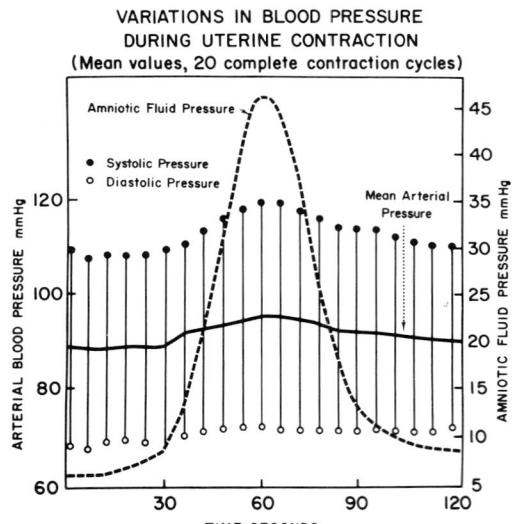

Figure III-7. The mean arterial pressure is somewhat elevated during the active phase of contractions.[3]

cycle (Fig. 8). We have measured increments as high as 40 mm Hg. in the femoral venous pressure. The increments are not always large. For example, in Figure 9 there is shown a segment of the record from the same patient as that of the previous illustration. At the lefthand portion of the illustration the patient was supine, and the rise of femoral venous pressure in response to uterine contractions was much less than in the previous illustration. When

uterine contractions

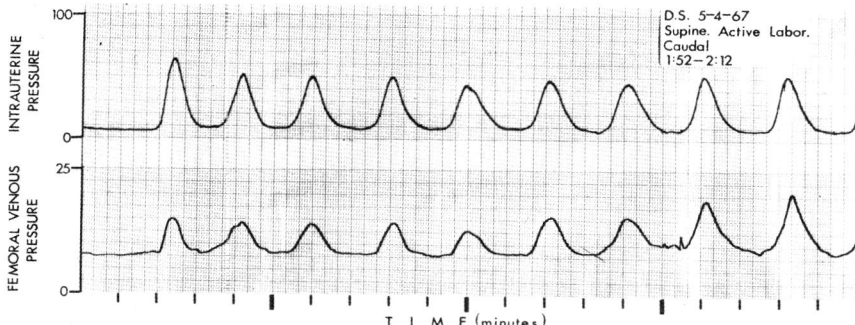

Figure III-8. In this tracing the intrauterine pressure change is reflected qualitatively but not quantitatively by pressure waves in the femoral venous pressure.

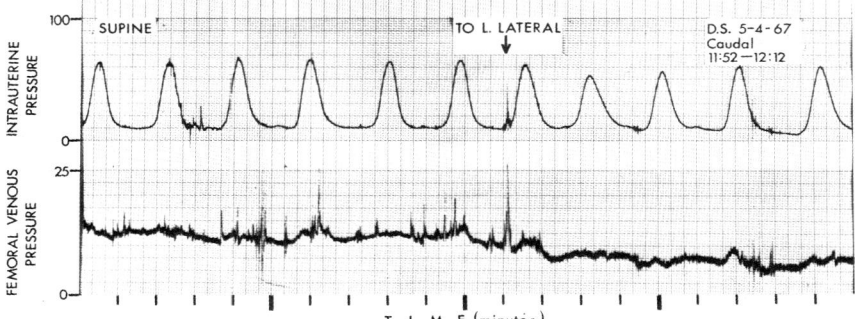

Figure III-9. Same patient as in Figure 8. In this portion of the record the response of femoral venous pressure to uterine contractions is much smaller. Some response is present even with the patient in a lateral position.

the patient turned to the lateral position, the response of the femoral venous pressure to contraction cycles diminished much further.

To the extent that the blood in the uterine circulation may differ from that of the extrauterine circulation, any such differences may be reflected in the general circulation after a large component of uterine blood has been rapidly extruded into the central circulation at the onset of a contraction. The alteration in arteriovenous oxygen difference as described by Palmer and

prenatal life

Walker has already been pointed out. These authors attributed the increase to the addition of more reduced blood from the uterus.

Another illustration may be furnished by blood lactic acid levels. If the lactic acid level within the uterus happens for any reason to exceed that in the extrauterine circulation, the peripheral arterial blood lactate levels may be altered in response to uterine contractions.[10] In Figure 10 are shown small elevations in brachial arterial blood lactate which appear to rise and fall in waves that bear a consistent relationship to the uterine contraction cycle. The lag time between the uterine pressure elevation and the blood lactate elevation is interpreted as a simple reflection of the subject's circulation time. A further expression of differences between the intrauterine and extrauterine characteristics is the factor of temperature. We have found that the brachial arterial blood temperature rises in association with a uterine contraction, the time sequence being approximately the same as that in the previous illustration involving lactic acid.[11]

Symptomatically the response of the normotensive patient to these cardiovascular changes is minimal. The only instance that I have encountered of a symptomatic response to these changes in the normotensive patient is occasional nasal congestion at the onset of a uterine contraction, usually in patients who already have some degree of nasal congestion.

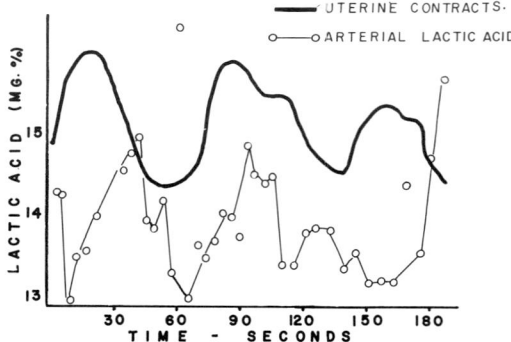

Figure III-10. Lactic acid determinations of blood taken from the brachial artery of a woman in active labor show cyclic alterations which bear some resemblance to the shape of uterine contractions.[10]

The Mechanical Effect of the Contraction on the Lower Circulation: an Added Factor

Let us now consider these mechanical phenomena associated with the contractions which are not circulatory in origin, but which have implications for the maternal circulation.

It has long been accepted as fact that when the pregnant patient is supine there may be some obstruction to venous return by the vena caval route.[12] The extensive studies of Kerr and Scott[13, 14] have further documented the phenomenon of the inferior vena caval syndrome. There has been considerable speculation as to whether or not the occasional large rise in femoral venous pressure is due primarily to caval compression or whether it may instead reflect the denial of entry into the central circulation by the blood being squeezed out of the uterus.

Evidence has also been brought forward indicating that the mechanical effect of a uterine contraction may compromise the adequacy of the lower

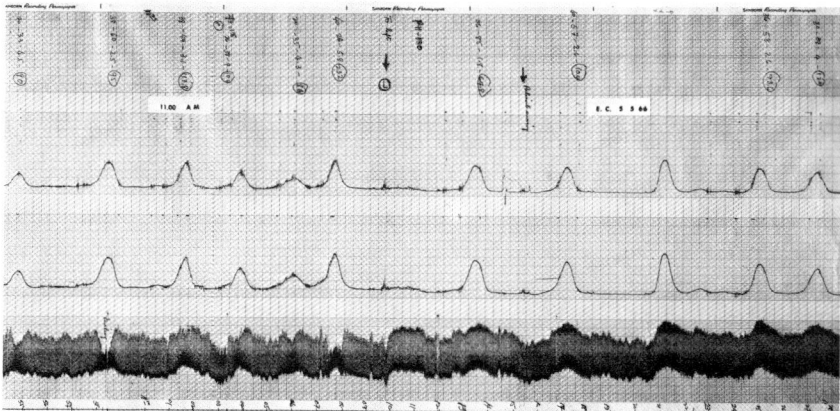

Figure III-11. A thirty-five-minute segment of a patient in spontaneous labor. The intrauterine pressure is being recorded (from two catheters) as well as the femoral arterial blood pressure. At the left portion of the record, with the patient supine, there are several sharp dips in the arterial blood pressure coincident with uterine contractions (Poseiro effect). In the midportion of the record the patient turns to her left side, and the Poseiro effect promptly disappears.

prenatal life

arterial circulation due to direct compression upon the vessels involved. Some time ago it was observed by Poseiro that under some circumstances the femoral arterial pressure would diminish greatly in association with a uterine contraction cycle.[15] This reduction in femoral pressure is in almost direct mirror image to the smooth ascent and descent of intrauterine pressure (Fig. 11). Inasmuch as the compressive effect which could affect the femoral arterial flow is exerted at a level where the uterine circulation also may be

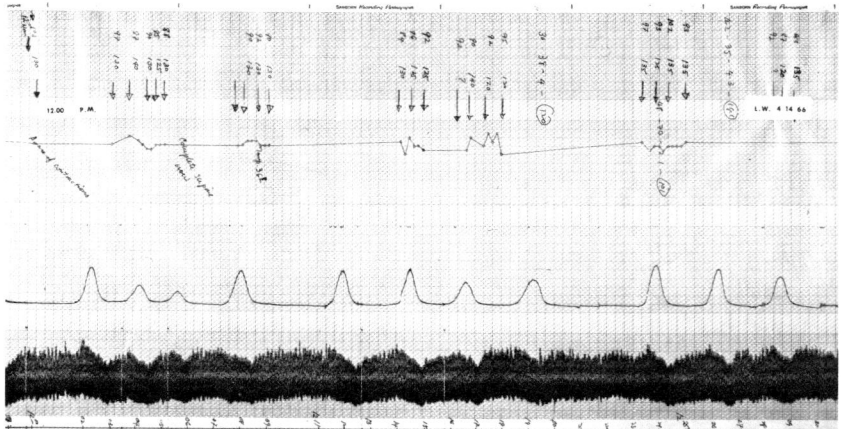

Figure III-12. An unusual manifestation of the Poseiro effect is seen in this tracing of a woman in early labor, lying quietly supine. The reduction in arterial blood pressure is relatively slow and reaches its maximum only during the terminal phase of the contraction cycle.

compromised, the assumption has been (and there is some evidence to support this) that the appearance of the Poseiro effect with the patient in the supine position indicates that the uterine circulation, at least on one side, is most inadequate during a contraction cycle. More recently Bieniarz[16] has extended these observations and brought forth additional evidence indicating that during some uterine contractions, with the patient in the supine position, the circulation through the aorta itself may be seriously impeded.

It may be noted in Figure 11 that with the onset of each contraction the femoral arterial blood pressure diminishes and returns to its precontraction state only upon the completion of the contraction. At the point where the patient turned to her left, however, the Poseiro effect disappears, and there is a small rise in the blood pressure in association with each contraction. This illustrates the general truth that when such a situation is present it can be cleared up very rapidly either by shifting the uterus to the left or turning the patient to her side, usually toward the left.

An unusual variant of the Poseiro effect appears in Figure 12. In this instance, for unknown reasons, the diminishment of arterial blood pressure in response to uterine contractions has its onset only after the contraction is well underway, while complete recovery to the precontraction state appears only after the contraction has been completed.

What is the significance of the Poseiro effect with or without evidence of aortic compression? Fortunately this is a phenomenon which is observed only in unusual instances. It can sometimes be induced or released by very small shifts in the position of the uterus, or of the patient, or of the patient's leg. The diminished arterial blood pressure occurs only at a time when the uterus is not being actively perfused anyway; i.e., almost always during the active phase of the contraction cycle. Thus, if it is not associated with supine hypotension, the likelihood of its doing much damage is very small. If we do not wish to entertain the thought that patients might develop the Poseiro effect during labor, however, all we have to do is to permit women to labor primarily on their sides rather than in the supine position.

Relationship of Uterine Contractions to Hypotension

All too commonly we observe hypotension in the parturient. Especially in early labor, the woman may be experiencing the supine hypotensive syndrome of pregnancy. A tendency to the development of this state may be accentuated by prolonged dehydration (it occurs much more frequently in hot weather, in the dehydrated patient, and after prolonged use of diuretic agents). Late in labor a frequent cause of hypotension is the administration of conduction analgesia, either spinal, caudal, or epidural, with the woman kept in the supine position. The prime factor in most supine hypotension observed during labor probably is partial obstruction of venous return from

prenatal life

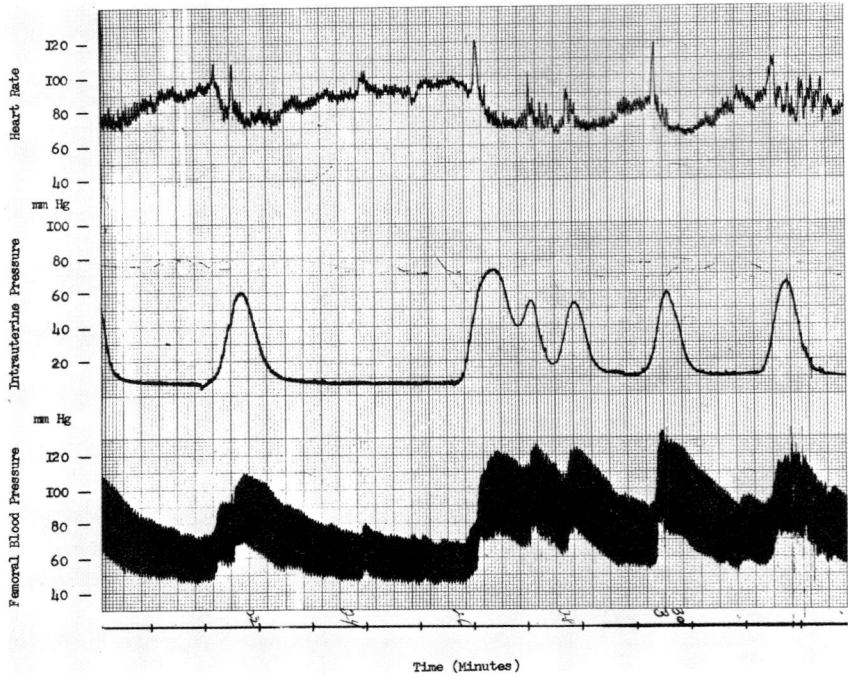

Figure III-13. The heart rate, intrauterine pressure, and femoral blood pressure are being recorded in a patient who was exhibiting supine hypotension. The blood pressure makes a dramatic response to the onset of each uterine contraction. The response continues after the contraction is finished. (From C. H. Hendricks: Technique and clinical significance of amniotic fluid pressure recording. *Clin Obstet Gynec*, 9: 535-553, 1966.)

the lower extremities, which in turn has the effect of reducing the effective circulating volume. The process can be accentuated by the peripheral vasomotor block which occurs below the level of conduction anesthesia.[17] It is further abetted by any medication or condition which may contribute to the development of any degree of hypovolemia.

Clinically significant degrees of hypotension can be very severe and yet go unrecognized. It is not at all uncommon after the administration of conduction analgesia and before hypotension has been documented to have

uterine contractions

the patient first complain of thirst, then turn pale, begin to sweat, feel nausea, or suddenly feel drowsy or faint. Because of the well-known effect of contractions upon the blood pressure levels, it is mandatory that such a patient should have her blood pressure recorded in the interval between contractions, rather than with contractions, if one wishes to document her true blood pressure values.

These symptoms of impending shock are usually relieved by the onset of the next contraction. One frequently observes that the patient who becomes drowsy between contractions becomes alert and feels quite well during the active contraction phase. Let us see why this should be so. Figure 13 shows the record of a woman in active labor who was having contractions

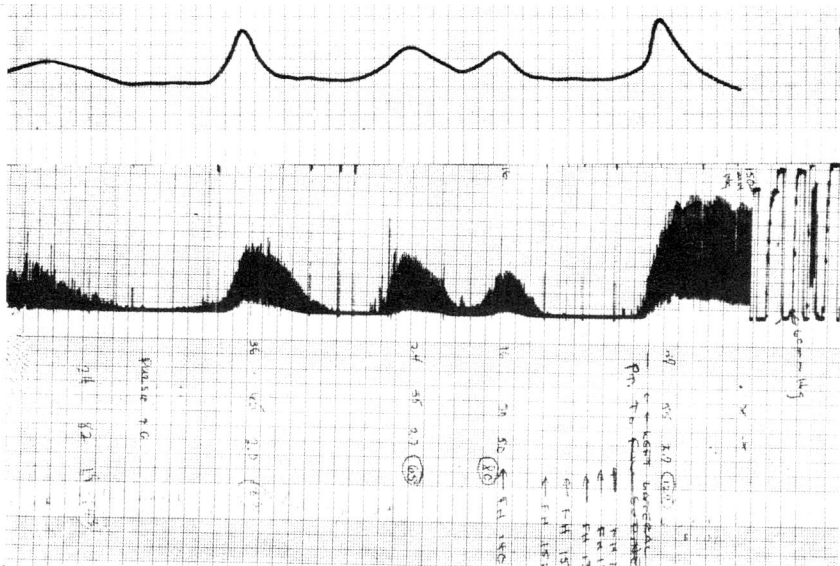

Figure III-14. Intrauterine pressure and brachial arterial pressure are being recorded in a woman with supine hypotension. The arterial pressures are being recorded only in excess of 60 mm Hg. The severe hypotension is temporarily relieved by the presence of a uterine contraction. When the patient turns to her left side the hypotension is dramatically relieved.

prenatal life

in an irregular pattern. Between contractions the uterine activity was dropping to shock levels (below 80 systolic). Within seconds after the beginning of each contraction the blood pressure began a substantial rise, in some instances increasing the systolic pressure by as much as 50 mm Hg. and the diastolic pressure by as much as 25 mm Hg. After the contractions subsided the blood pressure again drifted slowly back to shock levels, subsiding much more slowly than did the uterine contraction wave, but continuing its downward drift until the onset of the next contraction, when the blood pressure would again be rapidly restored to levels within the normotensive range.

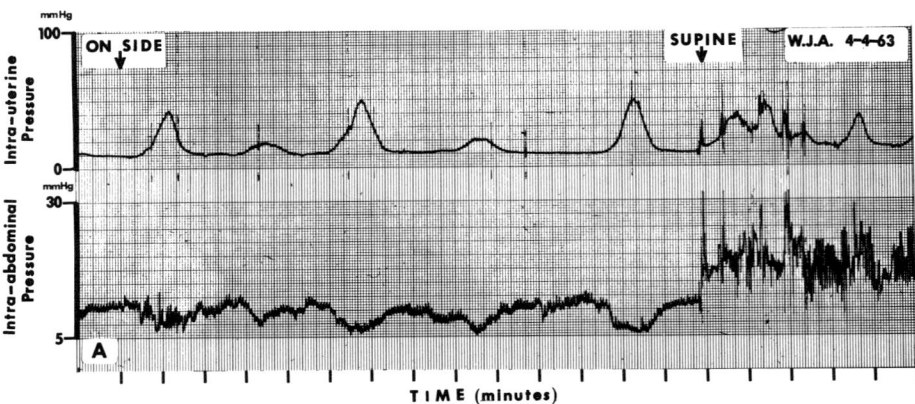

Figure III-15. Intraabdominal pressure responds slightly to alterations in uterine activity in this tracing. This is not a constant finding.

A more dramatic illustration of this phenomenon is shown in Figure 15 where the brachial arterial blood pressure was being recorded. Only blood pressures above 60 mm Hg. were being recorded. As long as the patient remained in the supine position the systolic blood pressure exceeded 60 mm Hg. only in association with uterine contractions. The blood pressure tracings produced a pattern which closely mimicked the basic pattern of the contractions. Near the end of the tracing, the patient turned from supine to her left side, and the blood pressure was rapidly restored to full normotensive levels.

Why should uterine contractions have such a salutary effect upon the blood pressure of a pregnant woman who is experiencing supine hypotension? Presumably this is the "pump-priming effect," wherein the rapid extrusion of blood from the uterus back into the central circulation restores the effective circulating volume to the range within which a normotensive pressure can be maintained. It is gratifying to observe that the pump-priming provides a beneficial effect on the circulation which outlasts the duration of the contraction itself.

This brings us to a point of potential clinical usefulness. In the unusual situation wherein the contractions are so infrequent that severe hypotension develops for too long a period after each contraction, the use of an oxytocin infusion within physiologic limits may be employed to "prime the pump" more frequently and thus to keep the patient's blood pressure above shock levels. It should be pointed out, however, that neither the mother nor the fetus in utero benefit from the conduct of any portion of her labor in the presence of any significant degree of hypotension. While the utilization of an oxytocin infusion at physiologic levels might be justified in exceptional cases, the more logical approach is to remove the major causes of the deficit in the effective circulating volume by (1) turning the patient on her side to relieve the obstruction to venous return from her lower extremities, (2) elevating her legs to drain a portion of the blood sequestered there back into the central circulation; and (3) administering very rapidly a salt-containing electrolyte solution (physiologic saline, Ringer's lactate solution, etc.), usually 500 to 1,500 cc.

One more comment about the pump-priming effect. It is a curious fact that the uterus, whose circulation we are dedicated to protecting in the interest of the fetus it contains, is capable under stress of exerting a corrective influence upon a failing effective circulating volume. Unfortunately such a failing central circulation is too often brought about in some measure by well-meaning, but misguided, medical measures. Fortunately, the condition can usually be relieved simply by proper postural management of the patient.

Effect of Contractions upon Abdominal Pressure

Figure 15 illustrates some changes in intra-abdominal pressure in response to uterine contractions. In this instance there was a decline of 2 to 3 mm Hg in response to each contraction. This change is not a constant one. The

prenatal life

response to contractions is much smaller than the approximately 7 mm rise in abdominal pressure induced simply by changing the patient to the supine position. The latter alteration is believed to be due principally to the fact that this postural change raises the uppermost portion of the uterus to a higher level, although other factors may be involved.

Effect of Contractions upon Cerebrospinal Fluid Pressure

Intrauterine pressure, arterial blood pressure, and cerebrospinal fluid pressure (CSFP) all appear to be interrelated.[18] As noted previously, the arterial blood pressure begins its rise almost simultaneously with the onset of the contraction. The CSFP also rises in response to a uterine contraction, but it does not begin its increase until a few seconds after the elevation in blood pressure has become apparent (Fig. 16). Another interesting, but unexplained, feature of its behavior is that the wave of increased CSFP subsides more slowly than either that of the uterine contraction or that of the blood pressure (Fig. 17). The total rise in the CSFP is small, usually within the range of 1.5 to 8 mm Hg.

The general rule is that the rise in the CSFP may be observed in each instance where there is a significant rise in the arterial blood pressure. Contrary to the observations of others,[19] the wave of response occurs even in painless contractions (Fig. 18). During painful contractions one observes the same basic response with superimposed irregularities induced by the patient's movement and other expressions of her discomfort (Fig. 19).

The question frequently arises as to whether or not intrathecal medication may be administered during a contraction. Hesitation upon this point is apparently based partially upon the erroneous impression that the CSFP rises greatly with each contraction, and that this pressure somehow alters the flow characteristics within the spinal canal. There is no evidence, though, that a uterine contraction per se ever does either of these things. Kretchmer and Vasicka[20] agree with our impression that there is no specific interdiction to giving spinal anesthesia during a contraction. Another objection raised to such a practice is that movements of the patient in hard labor may be harder to control during a contraction: this argument is more cogent, but discussion of it is beyond the scope of the present paper.

uterine contractions

Figure III-16. During two contractions, the intrauterine pressure, cerebrospinal fluid pressure (CSFP), and arterial blood pressure are recorded at a slow rate (in the left part of the illustration) and at a much more rapid rate (in the right part of the illustration). Note the consistent response of the CSFP to the uterine contraction.[18]

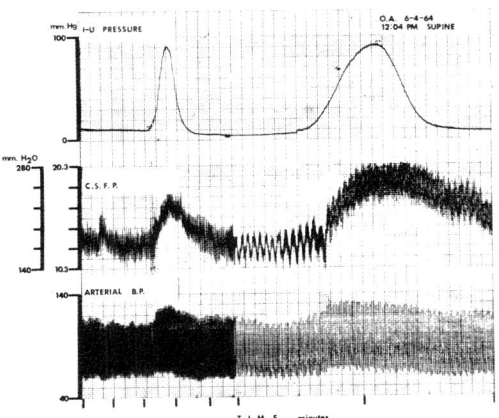

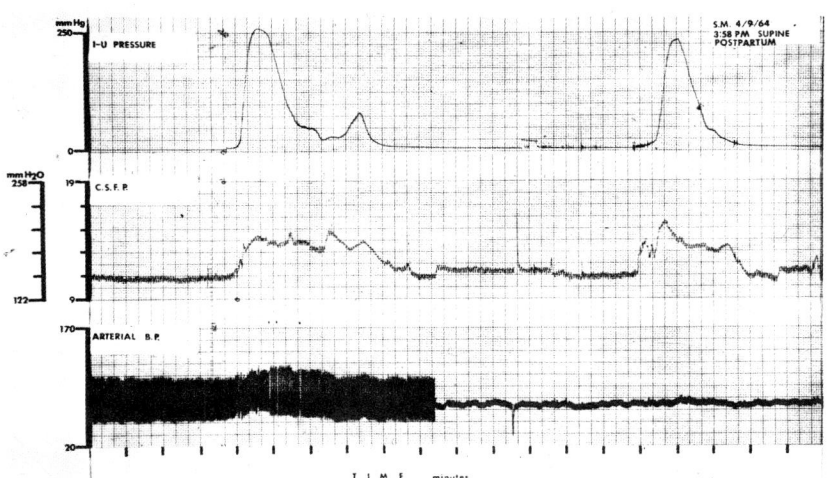

Figure III-17. The CSFP response to the uterine contraction tends to last longer than the contraction itself.[18]

prenatal life

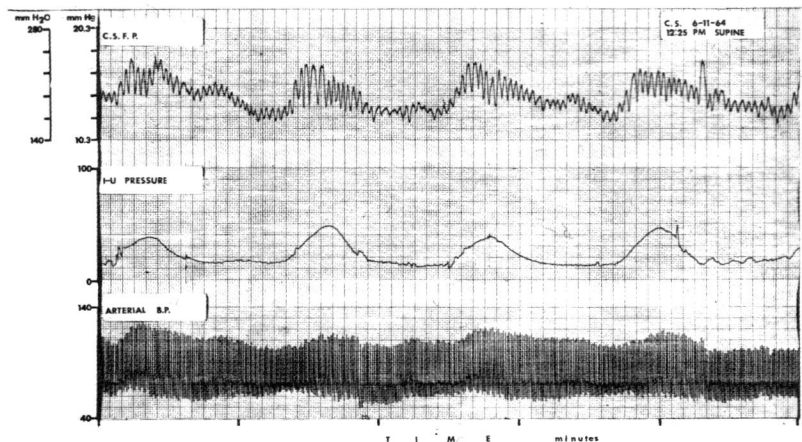

Figure III-18. The rise in CSFP occurs even in patient who is not aware that she is having contractions.[18]

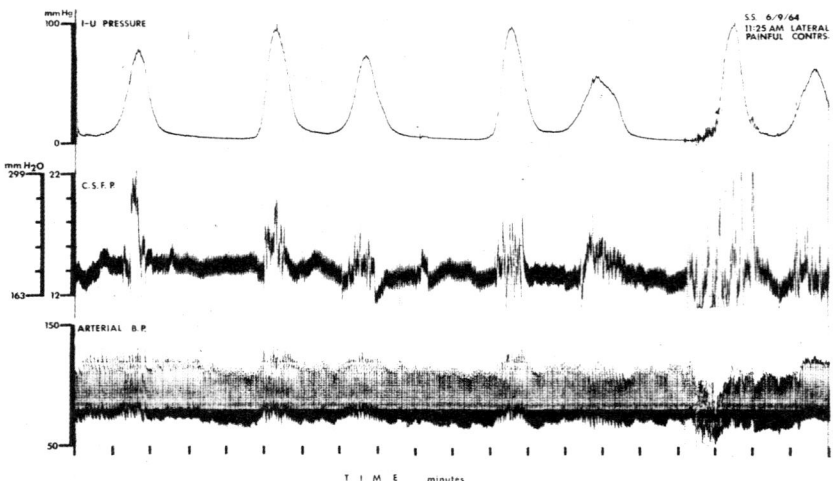

Figure III-19. With painful contractions the CSFP (middle line) still rises with contractions, but there are superimposed irregularities brought about by the patient's discomfort and motion.[18]

Summary

The uterus at term is an organ of imposing magnitude whose actions produce major measurable effects upon the mother who provides it shelter. Like the heart, the uterus is a pump. It is elegantly designed to expel the fetus at term, through its pumping action.

During its contraction cycle, like the heart, it also pumps out a substantial quota of blood, what we might call the "uterine output" being equivalent to a "stroke volume" of 250 to 300 cc. In this huge organ the cyclically repeated extrusion of blood, attrition of blood input, and subsequent resumption of full perfusion impose a challenge upon the adaptive powers of the circulatory system.

Each "stroke" of this peripheral "heart" puts out pressure, heat, and blood whose chemical nature has been altered by its trip through the uterus. The successful disposition or dispersal of the extruded products is a vital part of the birth process.

References

1. Palmer, A. J., and Walker, A. H. C.: The maternal circulation in normal pregnancy. *J Obstet Gynaec Brit Comm, 56:* 537-547, 1949.
2. Ramsey, E. M., Martin, C. B., Jr., and Donner, M. W.: Radiographic studies of the venous drainage of the placenta in rhesus monkeys. *Obstet Gynec, 25:* 417-418, 1965.
3. Hendricks, C. H.: The hemodynamics of a uterine contraction. *Amer J Obstet Gynec, 76:* 969-982, 1958.
4. Caldeyro-Barcia, R.: In *Physiology of Prematurity*, ed. J. T. Lanman, New York, Josiah Macy, Jr. Foundation, 1957, p. 128.
5. Woodbury, R. A., Hamilton, W. F., and Torpin, R.: The relationship between abdominal uterine and arterial pressures during labor. *Amer J Physiol, 121:* 640-649, 1938.
6. Borell, U., Fernström, I., Ohlson, L., and Wiqvist, N.: Effect of uterine contractions on the human uteroplacental blood circulation. *Amer J Obstet Gynec, 89:* 881-890, 1964.
7. Ramsey, E. M., and Harris, J. W. S.: Comparison of uteroplacental vasculature and circulation in the rhesus monkey and man. *Carnegie Contrib Embryol, 38:* 59-70, 1966.
8. Burwell, C. S.: Circulatory adjustments to pregnancy. *Bull Hopkins Hosp, 95:* 115-129, 1954.

9. Remington, J. W., Noback, C. R., Hamilton, W. F., and Gold, J. J.: Volume elasticity characteristics of human aorta and prediction of stroke volume from pulse pressure. *Amer J Physiol, 153:* 298-308, 1948.
10. Hendricks, C. H.: Studies on lactic acid metabolism in pregnancy and labor. *Amer J Obstet Gynec, 73:* 492-506, 1957.
11. Hendricks, C. H.: Unpublished data.
12. Howard, B. K., Goodson, J. H., and Mengert, W. F.: Supine hypotensive syndrome in late pregnancy. *Obstet Gynec, 1:* 371-377, 1953.
13. Scott, D. B., and Kerr, M. G.: Inferior vena caval pressure in late pregnancy. *J Obstet Gynaec Brit Comm, 70:* 1044-1949, 1963.
14. Kerr, M. G.: The mechanical effects of the gravid uterus in late pregnancy. *J Obstet Gynaec Brit Comm, 72:* 513-529, 1965.
15. Poseiro, J. J., Massi, G. B., and Bieniarz, J.: Hipotension arterial regional causada por la contracción uterina. IV. Congr. Uruguayo de Ginecotocol. Montevideo, 1964, vol. II, p. 925.
16. Bieniarz, J., Maqueda, E., and Caldeyro-Barcia, R.: Compression of aorta by the uterus in late human pregnancy. *Amer J Obstet Gynec 95:* 795-808, 1966.
17. Lull, C. B., and Hingson, R. A.: *Control of Pain in Childbirth.* 3d ed., Philadelphia, Lippencott, 1948, p. 122.
18. Hopkins, E. L., Hendricks, C. H., and Cibils, L. A.: Cerebrospinal fluid pressure in labor. *Amer J Obstet Gynec, 93:* 907-916, 1965.
19. Marx, G. F., Zemaitis, M. T., and Orkin, L. R.: Cerebrospinal fluid pressures during labor and obstetrical anethesia. *Anesthesiology, 22:* 348-354, 1961.
20. Vasicka, A., Kretchmer, H., and Lawas, F.: Cerebrospinal fluid pressures during labor. *Amer J Obstet Gynec, 84:* 206-212, 1962.

chapter IV

Adrenergic Mechanisms in Human Myometrial Control

Richard W. Stander, M.D. and Tom P. Barden, M.D.

In 1948, Ahlquist reported upon a series of observations concerning the effects of a variety of phenethanolamines upon smooth muscle obtained from various organs in several species of animals.[1] At this time he proposed the concept that two types of receptors were to be found in smooth muscle. The alpha receptor, when excited or stimulated, resulted in smooth muscle contraction in most organ systems, while stimulation of beta receptors resulted in inhibition of smooth muscle activity. It became apparent that some of the sympathomimetic drugs, such as norepinephrine, are primarily active at alpha receptor sites and stimulate contractions in most smooth muscle. Others, as exemplified by isoproterenol, exert their influence at beta receptors and result in inhibition of smooth muscle activity. Some sympathomimetic drugs, such as epinephrine, are capable of stimulation of both alpha and beta receptors. These relationships are diagramatically represented in Figure 1.

In defining the final results (stimulation or inhibition) of sympathomimetic agents capable of activating both alpha and beta receptor sites Ahlquist felt that the relative population densities of each receptor type in a given organ of a certain species were important determinants.

The concept of alpha and beta receptor sites is a functional rather than an anatomic one, since specific receptor sites are not demonstrable by

From the Departments of Obstetrics and Gynecology, University of Cincinnati Medical Center and Indiana University Medical Center.

prenatal life

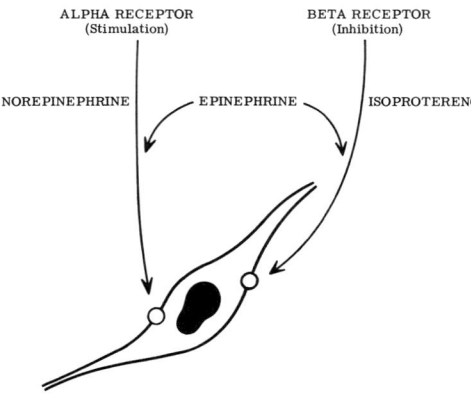

Figure IV-1. Diagrammatic representation of alpha and beta adrenergic receptors in smooth muscle cell. Norepinephrine is primarily a stimulant to the alpha receptors and stimulates smooth muscle contraction. Isoproterenol affects beta receptors and thereby inhibits smooth muscle contractility. Epinephrine is capable of simultaneous stimulation of both alpha and beta adrenergic receptors.

any techniques used to delineate other receptor sites, i.e., loci of antigen-antibody reactions on the red blood cell envelope.

Further studies of effects of sympathomimetic agents upon alpha and beta receptor sites were enhanced by discovery of agents capable of specific blockade of one type of receptor or the other. Among compounds known to block the alpha adrenergic receptors are those of the imidazoline group.[2] In 1958, Powell and Slater reported the discovery of the first compound capable of specific beta receptor blockade, dichloroisoproterenol.[3] However, this compound was found to have intrinsic sympathomimetic activity in many test situations, thus limiting its experimental usefulness. Newer beta blocking agents have since become available with few or none of the undesirable features of dichloroisoproterenol. Among these is propranolol (Inderal*).

Use of specific blocking agents has allowed more accurate evaluation of sympathomimetic amines upon alpha and beta adrenergic receptor sites through the selective ability to block one receptor site while the other receptor site remains unaffected and responsive. Since activation of both receptor sites results in diametrically opposed actions, inhibition vs. stimulation, the alpha or beta receptor activity of any compound can be theoretically studied as an isolated mechanism by blockade of the opposing receptor. This not only pro-

* Ayerst Laboratories, New York.

adrenergic mechanisms and myometrial control

vides a method for evaluating effects of exogenous sympathomimetic amines by in vivo as well as in vitro experiments, but provides a method of studying the effects of endogenous catecholamines upon specific organs in the intact animal or in man. Figure 2 is a schematic representation of the mechanism of specific adrenergic receptor blockade in smooth muscle.

Many questions concerning regulation of human myometrial function during pregnancy remain unanswered. The role of endogenous catecholamines in stimulation or inhibition of human gestational myometrium has not been systematically explored. Ahlquist's conception of alpha and beta receptor sites has not been related to human myometrium by a combined approach of in vitro and in vivo studies or the employment of blocking agents to isolate catecholamine effects to either alpha or beta receptor activity. This hiatus in our knowledge prompted the following studies.

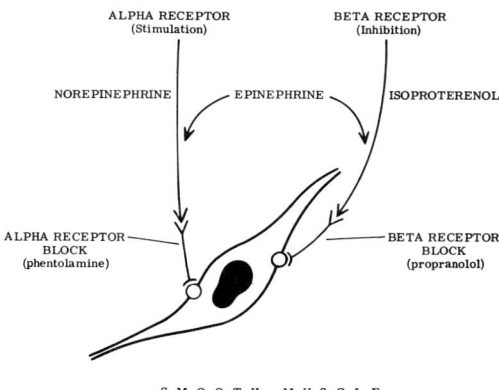

Figure IV-2. Diagrammatic representation of alpha and beta adrenergic receptor blockade in smooth muscle cell. Phentolamine, a representative of the imidazoline group, has a strong alpha receptor blocking component while propranolol acts as a blocker at beta receptor sites.

Materials and Methods
In Vitro Studies

Specimens of human gestational myometrial were obtained at the time of cesarean section from the lower uterine segment. All samples from gravid uteri were obtained during the last month of gestation or during labor. In vitro studies were carried out immediately in some cases. In others, speci-

prenatal life

mens were stored in normal saline at 3° C for periods up to forty-eight hours prior to testing.

Individual preparations of myometrium measuring $4 \times .5 \times .5$ cm were made from each sample. Measurements were obtained in the resting state. In most instances two specimens from the same sample were run simultaneously to provide suitable controls for the experimental preparations. Each preparation was suspended in a muscle bath chamber of 75 ml capacity. The preparation was fixed to the bottom of the chamber, while the top was suspended from a force-displacement transducer, allowing measurement of isometric tension. A multichannel recorder made a permanent record of isometric tension after suitable preamplification of the signal.

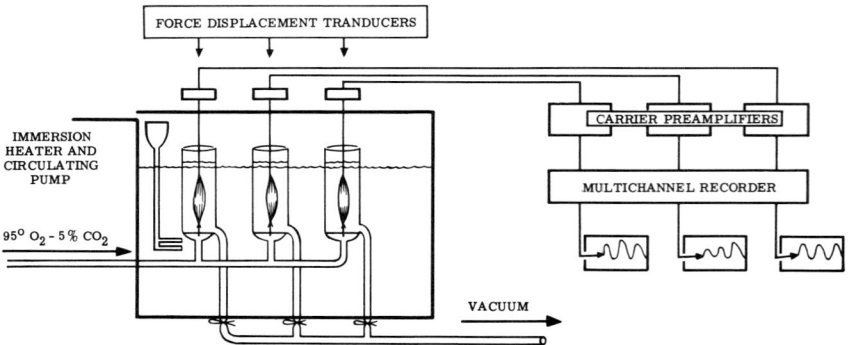

Figure IV-3. Diagrammatic representation of method of performing in vitro pharmacologic studies upon isolated specimens of human gestational myometrium.

The muscle chambers have fritted discs in the bottom allowing a continuous bubbling of a mixture of 95 percent O_2 and 5 percent CO_2 throughout the nutrient medium surrounding the myometrial sample. The chambers are also connected to a vacuum system allowing for rapid emptying. All chambers are immersed in a tank with a thermostatically-controlled heater-circulator designed to keep muscle bath temperature at a constant 37.5° C ($\pm$ 0.2°). The apparatus for in vitro studies is depicted in Figure 3.

The nutrient medium used in the muscle chambers is a modified mammalian Krebs' solution which has been previously described.[4]

adrenergic mechanisms and myometrial control

In Vivo Studies

Such studies were undertaken in a unit especially equipped for pharmacologic studies during human pregnancy. Studies were carried out on subjects who were in early labor or who were close enough to term to allow safe induction of labor.

To allow measurement and continuous recording of true intrauterine pressure, the method of transabdominal pressure measurement developed by Alvarez and Caldeyro-Barcia was used.[5] In some subjects transabdominal amniocentesis for the purpose of measuring intrauterine pressure was not feasible; in such cases a continuous recording of estimated intrauterine pressure employing the Malmström transducer was used.[6] Maternal digital systolic blood pressure was often continuously recorded by the use of a finger microphone and an occlusive cuff. Pressure in the latter is controlled by solenoid switching of the pump mechanism by signals from the microphone. In some studies simple repetitive auscultation of the indirect brachial systolic and diastolic blood pressure was manually recorded.

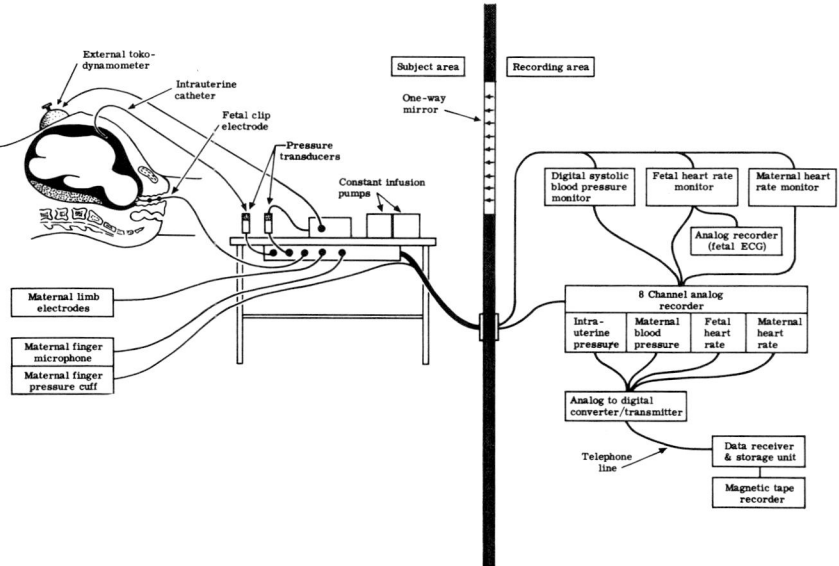

Figure IV-4. Diagrammatic representation of unit for studying material-fetal pharmacologic responses in advanced human pregnancy.

79

prenatal life

Maternal limb leads were provided for intermittent recording of the maternal ECG when such recording was deemed desirable or for continuous recording of maternal heart rate. In studies of fetal cardiac response of maternal drug infusion an electrode was attached to the presenting part of the fetus and the isolated, modified, and amplified fetal R wave was used to trigger a linear cardiotachometer providing a continuous recording of fetal heart rate on a beat-to-beat basis.[7]

All pharmacologic agents were given at controlled infusion rates by constant-infusion pumps.

The recording area is physically separate from the area housing transducers and subject. It contains equipment for analog recording, but in addition, has an analog-to-digital converter/transmitter providing for storage of physiologic data in digital form at a remote digital tape drive. The subject area is under constant surveillance from the recording area by means of a one-way mirror and intercom system. Figure 4 is a diagrammatic illustration of subject and recording areas in use during these studies.

Results

Initial in vitro testing was carried out on human gestational myometrial preparations to determine whether either the alpha receptor blocker phentolamine (Regitine*), or the beta receptor blocker, propranol (Inderal**) has intrinsic activity resulting in alteration of spontaneous activity of the myometrial preparations. Both blocking agents were found to suppress spontaneous contractility in concentrations of 100 $\mu g/ml$. However, both were subsequently found to have significant specific receptor blocking activity at concentrations much lower than those associated with myometrical depression. In vitro testing of norepinephrine was carried out upon paired myometrial samples at concentrations ranging from .001 to 10 $\mu g/ml$. When the paired strips from the same myometrial sample had achieved a relatively stable level of spontaneous activity, one strip was treated with either the alpha blocker, phentolamine, or the adrenergic beta receptor blocking agent, propranol. Fifteen to twenty minutes later both strips were treated with identical concentrations of norepinephrine. These experiments indicated that the stimulatory action of norepinephrine upon isolated human gestational myometrium was

* Ciba Pharmaceutical Company, Summit, New Jersey.
** Ayerst Laboratories, New York, N.Y.

adrenergic mechanisms and myometrial control

mediated through adrenergic alpha receptors, since it could be completely or partially blocked by pretreatment of one preparation with phentolamine. Figure 5 depicts a characteristic response of isolated myometrial preparations to norepinephrine with and without preceding alpha receptor blockade. No adrenergic beta receptor activity of norepinephrine (inhibition of myometrial activity) could be demonstrated after alpha receptor blockade. These findings were supported by failure to record an enhanced response of spontaneous contractility to norepinephrine following beta receptor blockade by propranolol.

Experiments with epinephrine in vitro produced results similar to those obtained with norepinephrine. Irrespective of epinephrine concentration or type of concentration of adrenergic blocking agent used, no beta receptor (inhibitory) activity of epinephrine was evident in these studies. When

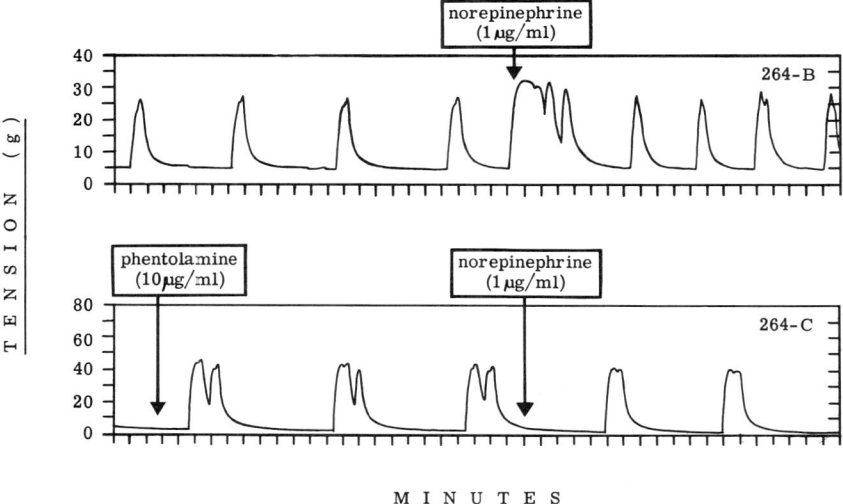

Figure IV-5. Isometric recording of spontaneous contractility of paired myometrial preparations. In 264-B (upper recording) a definite stimulatory response is noted to norepinephrine at a concentration of 1 μg/ml. However, there is no stimulatory response to an identical concentration of norepinephrine in the companion strip, 264-C which has been pretreated with phentolamine, 10 μg/ml.

prenatal life

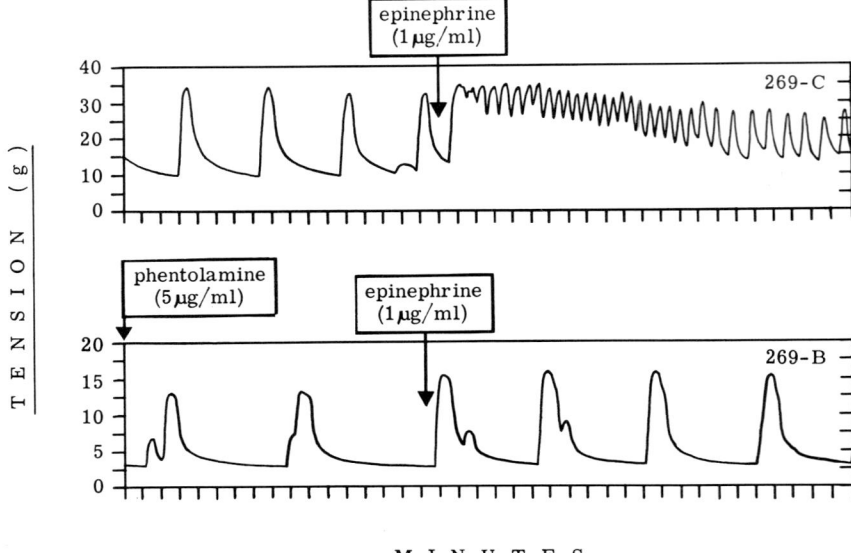

Figure IV-6. Isometric tension measurements of the effect of epinephrine, 1 μg/ml, upon spontaneous contractility of an isolated segment of human gestational myometrium (upper tracing). Pretreatment of a companion strip (269-B) with the alpha adrenergic receptor blocking agent, phentolamine, blocks the stimulatory response to a concentration of epinephrine identical to that used in the unblocked preparation.

alteration of spontaneous myometrial activity was achieved with epinephrine, it was always characterized by an increase in activity, notable in the frequency of contractions. Such stimulation could be completely or partially blocked by pretreatment of the preparation with phentolamine (Fig. 6).

It was, however, possible to demonstrate that adrenergic beta receptor activity did exist in isolated preparations of human myometrium. By using isoproterenol in concentrations ranging from 0.1 to 1.0 μg/ml definite inhibition of spontaneous contractility of isolated myometrial strips could be demonstrated. This was of a transient nature and primarily characterized by a reduction in the maximum tension developed during the contraction cycle

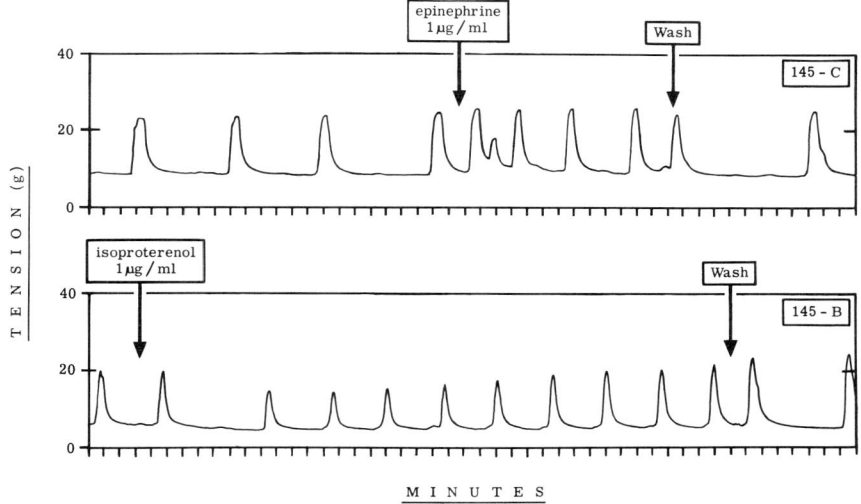

Figure IV-7. Comparative responses of equal concentrations of epinephrine and isoproterenol upon spontaneous contractility of paired strips of human gestational myometrium. Although epinephrine, 1 µg/ml, results in stimulation of the strip (upper tracing) a definite inhibition is noted following addition of isoproterenol to 145-B at a concentration of 1 µg/ml.

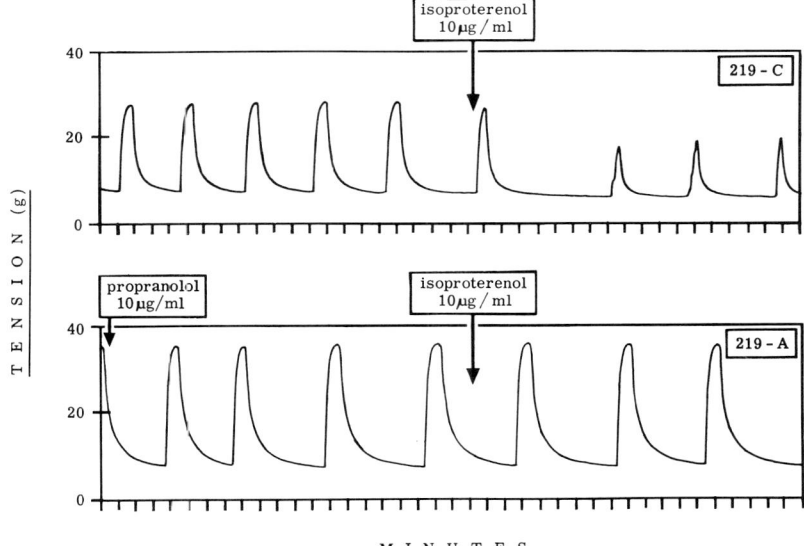

Figure IV-8. Upper tracing shows inhibition of spontaneous contractility of an isolated human gestational myometrial preparation to isoproterenol, 10 µg/ml. Pretreatment of the companion strip, 219-A, with the beta receptor blocking agent, propranolol, prevents the inhibitory response to a concentration of isoproterenol identical to that used on the unblocked preparation.

of the preparation (Fig. 7). Further evidence concerning the adrenergic beta receptor activity of isoproterenol on isolated myometrial preparations was provided by the observations that the beta receptor blocker, propranolol, reduces or eliminates the inhibition of the preparations by isoproterenol (Fig. 8).

Thus, it was established that alpha and beta adrenergic receptor activity could be demonstrated by alteration of contractility in isolated preparations of human gestational myometrium. It was also demonstrated that the alpha receptor blocking agent, phentolamine, and the beta receptor blocking agent, propranolol, were capable of specific receptor blockade at concentrations lower than those involving intrinsic inhibition.

With the knowledge that many discrepancies are known to exist between results obtained by in vitro testing and those observed during in vivo studies, observations were carried out to determine whether or not specific adrenergic receptor activity could be demonstrated in myometrial activity of the intact gravid human uterus.

To this end several volunteer human pregnant subjects, at or near term, were admitted to the study unit. Pregnancies were uncomplicated. Intrauterine pressure was measured or estimated as previously described. Maternal blood pressure was periodically obtained by indirect auscultation. Maternal heart rate was continuously recorded in all instances. In some subjects in whom artificial rupture of the membranes was feasible, fetal heart rate was continuously recorded. In subjects demonstrating low levels of spontaneous uterine contractility, uterine activity was frequently enhanced by continuous infusion of dilute infusions of synthetic oxytocin (Syntocinon*).

Intravenous infusion of norepinephrine into pregnant subjects at or near the term invariably increased the rate of uterine contractions when a critical infusion rate had been reached. This response appears to parallel that obtained by in vitro studies and confirms previous reports.[9] Furthermore, the stimulatory response of the intact human gravid uterus to norepinephrine could be greatly reduced or completely blocked by prior administration of the alpha blocker, phentolamine. A characteristic response of the intact gravid uterus to norepinephrine before and after administration of phentolamine is shown in Figure 9.

* Sandoz Pharmaceuticals, Hanover, New Jersey.

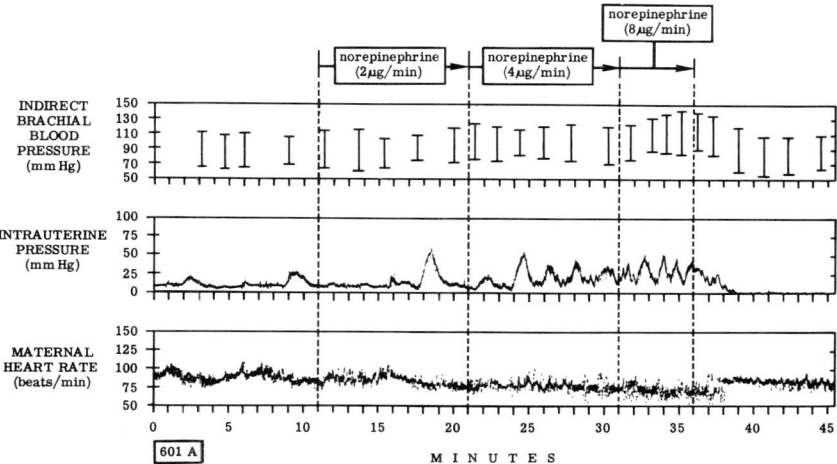

Figure IV-9a. Infusion of norepinephrine at increasing rates in human pregnant subject at term. A substantial increase in uterine activity (middle tracing) is noted at norepinephrine infusion rates of 4 and 8 μg/minute. Moderate alterations in maternal blood pressure also occur.

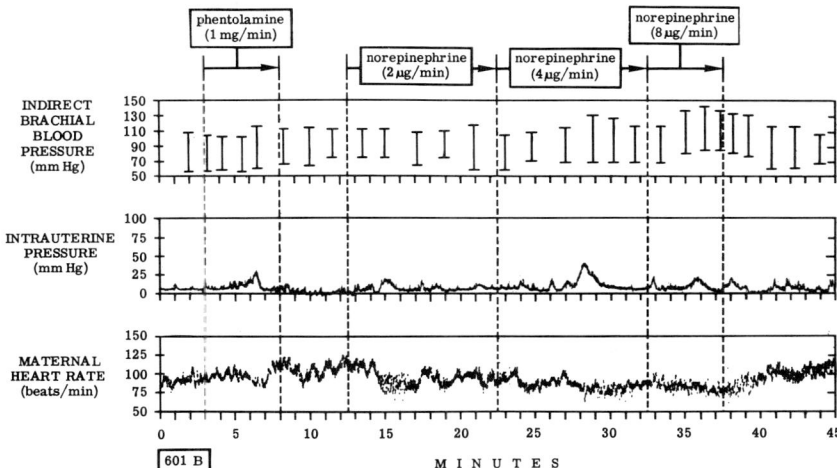

Figure IV-9b. Norepinephrine infusion is repeated following a five-minute infusion of the alpha receptor blocking agent phentolamine at a rate of 1 mg/min. A marked reduction in uterine response to norepinephrine infusion is noted. Persistence of maternal blood pressure elevation, especially at the 8 μg/min infusion rate, occurs despite phentolamine blockade.

prenatal life

With respect to epinephrine a disparity was noted between in vitro studies and in vivo studies. Infusion of epinephrine into the human pregnant subject generally resulted in a transient reduction of either spontaneous or induced uterine contractility, as previously reported by Stroup as well as Pose and his coworkers.[10, 11] This was in contrast to stimulation of isolated preparations by epinephrine. However, the inhibition of spontaneous and induced myometrial contractility of the intact gravid human uterus by epinephrine could not only be blocked by the beta receptor blocking agent, propranolol, but in most instances, epinephrine reversal took place, i.e., following beta receptor blockade by propranolol, i.e., alpha receptor activity of epinephrine was revealed as evidenced by an increase in uterine activity. Thus, only by in vivo testing was the ability of exogenous epinephrine to act at both alpha and beta adrenergic receptor sites in human gestational myometrium demonstrated. This phenomenon is illustrated in Figure 10.

Finally, the beta blocker propranolol was used in a fashion designed to probe the potential role of endogenous catecholamines in inhibition of myometrial function of the intact gravid uterus. It already had a demonstrated ability to block the uterine inhibitory action of exogenous epinephrine in the human pregnant subject. In addition, no intrinsic myometrial effect of propranolol had been detected during infusions in such subjects. It was therefore reasonable to assume that any increase in uterine activity during infusion of propranolol alone would in all likelihood result from ablation of myometrial inhibitory activity of endogenous epinephrine. Although this hypothesis has not been extensively tested in human pregnant subjects, initial results are indeed encouraging. Several instances of improvement in the pattern of spontaneous and induced uterine activity have been recorded during and following propranolol infusion.[12] Figure 11 demonstrates typical alteration in pattern of uterine contractility resulting from infusion of propanolol into a human pregnant subject.

Comments

In terms of actual catecholamine content, assay has indicated that human gestational myometrium is almost entirely devoid of measurable catecholamines suggesting absence of significant noradrenergic innervation.[13] This,

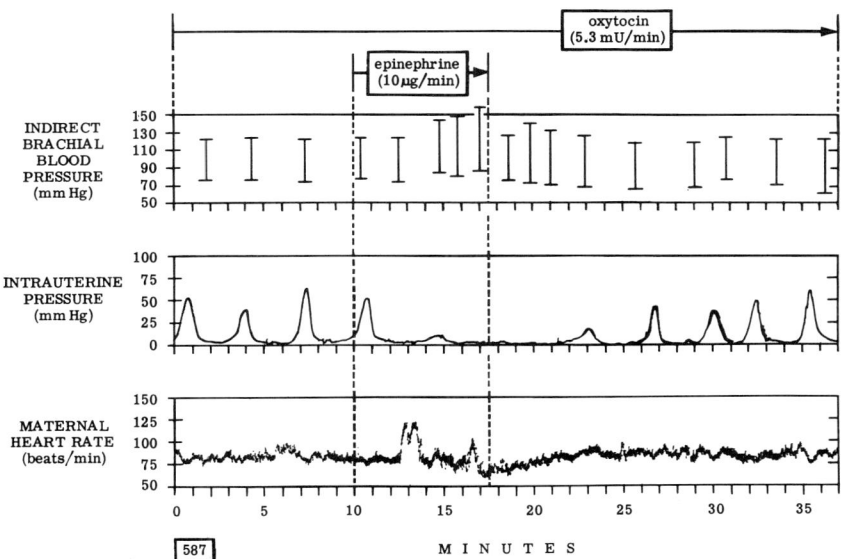

Figure IV-10a. Term pregnancy. Constant infusion of oxytocin at a rate of 5.3 mU/min. Marked inhibition of uterine contractility during and for a brief period following infusion of epinephrine at 10 µg/min. Modest elevation of maternal indirect brachial blood pressure is noted.

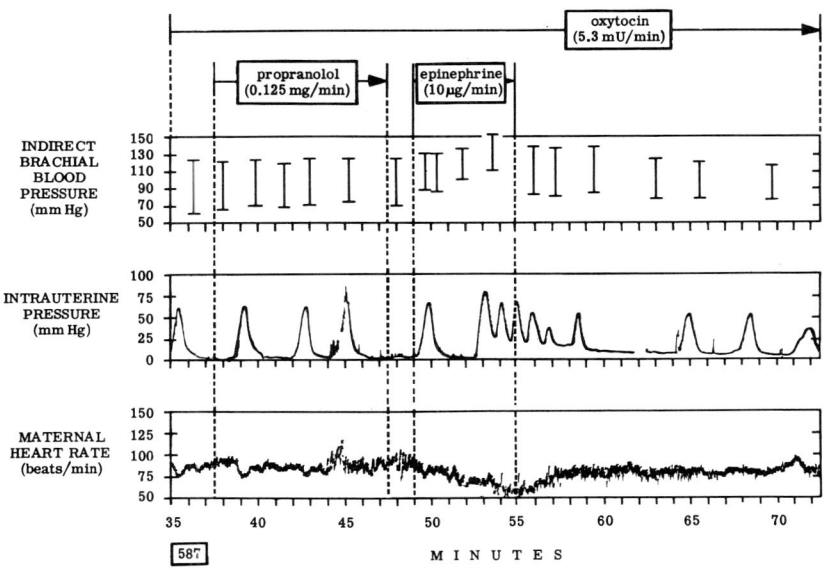

Figure IV-10b. Following ten-minute infusion of beta blocking agent, propranolol, at 0.125 mg/min, repeat infusion of epinephrine at 10 µg/min results in an increase rather than a decrease in uterine activity. Elevation of maternal diastolic blood pressure seems somewhat more pronounced as a result of epinephrine infusion following beta receptor blockage.

prenatal life

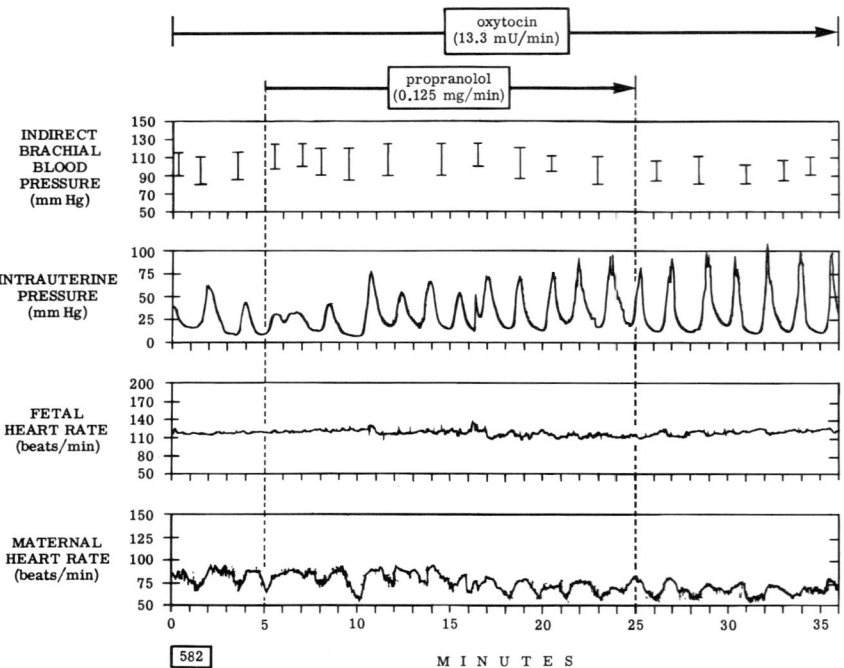

Figure IV-11. Plotting of indirect systolic and diastolic brachial blood pressure and continuous recording of intrauterine pressure, fetal heart rate and maternal heart rate in human pregnant subject. Infusion of oxytocin at 13.3 mU/min had resulted in relatively frequent contractions but of varying waveform and intensity. Concomitant infusion of propranolol at 0.125 mg/min for twenty minutes results in an increase in intensity of contractions as well as a more symmetrical pattern.

however, does not negate the possibility that catecholamines from other sources are capable of significantly altering myometrial activity.

Evidence already exists that abnormal labor patterns are associated with higher maternal levels of catecholamines than those found in uncomplicated labor patients.[14] Our own studies have confirmed previous findings that exogenous catecholamines can alter contractile patterns of human gestational myometrium both in vitro and in vivo. In addition, they have essentially substantiated Ahlquist's receptor theory with respect to human myometrium. Furthermore, a potentially important role of endogenous catecholamines in regulation of human gestational myometrial function has been underscored by the finding that specific receptor blockade often results in altered myometrial function. Evidence to date suggests that problems relating to adrenergic mechanism as factors in control of myometrial function are worthy of further elucidation.

References

1. Ahlquist, R. P.: A study of the adrenotropic receptors. *Amer J Physiol, 153:* 586, 1948.
2. Nickerson, M.: The pharmacology of adrenergic blockade. *Pharmacol Rev, 1:* 27, 1949.
3. Powell, C. E., and Slater, I. H.: Blocking of inhibitory adrenergic receptors by a dichloro analog of isoproterenol. *J Pharmacol Exp Ther, 122:* 480, 1958.
4. Stander, R. W., and Sherwood, E. A.: Spontaneous contractility of human myometrium in an isolated isometric system. *Amer J Obstet Gynec, 96:* 1060, 1966.
5. Alvarez, H., and Caldeyro-Barcia, R.: Contractility of human uterus evaluated by new methods. *Surg Gynec Obstet, 91:* 1, 1950.
6. Cordey, R., and Stander, R. W.: Estimation of intrauterine pressure by an external transducer. *Acta Obstet Gynec Scand, 43:* 115, 1964.
7. Stander, R. W., Barden, T. P., and Braunlin, R. J.: An instrument complex for intrapartum investigation of human fetal cardiac function. *Obstet Gynec, 22:* 265, 1963.
8. Stander, R. W., and Barden, T. P.: Adrenergic receptor activity of catecholamines in human gestational myometrium. *Obstet Gynec, 28:* 768, 1966.
9. Cibils, L. A., Pose, S. V., and Zuspan, R. P.: Effect of 1-norepinephrine infusion on uterine contractility and cardiovascular system. *Amer J Obstet Gynec, 84:* 307, 1952.
10. Stroup, P. E.: The influence of epinephrine on uterine contractility. *Amer J Obstet Gynec, 84:* 595, 1962.
11. Pose, S. V., Cibils, L. A., and Zuspan, F. P.: Effects of 1-epinephrine infusion on uterine contractility and cardiovascular system. *Amer J Obstet Gynec, 84:* 297, 1962.
12. Barden, T. P., and Stander, R. W.: Myometrial and cardiovascular effects of an adrenergic blocking drug in human pregnancy. *Amer J Obstet Gynec, 10:* 91, 1968.
13. Gaffney, T., Burket, R., and Warankow, S.: Catecholamine content of the pregnant and nonpregnant human uterus. *Obstet Gynec, 25:* 340, 1965.
14. Garcia, C. R., and Garcia, E. S.: Epinephrine-like substances in the blood and their relation to uterine inertia. *Amer J Obstet Gynec, 69:* 812, 1955.

chapter V

The Diagnosis and Treatment of Fetal Distress

Edward H. Hon, M.D.

In recent years attention has been focused on the problem of perinatal mortality and morbidity. With this increased interest the need for adequate methods for detection of fetal distress has become apparent, so that the present clinical criteria of fetal distress, which are based on stethoscopic sampling of fetal heart rate (FHR), are being reevaluated. The value of additional techniques such at fetal electrocardiography (FECG),[1-5] fetal phonocardiography (PCG),[6-7] and biochemical analysis of fetal scalp blood[8-11] have also been suggested as alternate or complementary methods for this purpose.

Several factors must be taken into account before judging the efficacy of a technique for the detection of fetal distress. Perinatal mortality figures, per se, may more accurately reflect the quality of newborn care rather than the quality of obstetric care. The Apgar score, widely used to evaluate neonatal condition at birth, is within itself subject to marked variations in interpretation. Followup studies based on infant growth and development, while more reliable, are complex and expensive and become quite difficult when carried out over a span of several years.

It is the purpose of this report to make observations on the use of FHR and fetal biochemistries as they were used in a recent study of 193 patients and to give a general appraisal of the value of these techniques for the diagnosis of fetal distress.

This investigation was supported in part by USPHS Research Grants HD 01467-04 and 5K03 HD 18295-07 from the National Institute of Child Health and Human Development.

prenatal life

Fetal Heart Rate Monitoring

When FHR is used for the evaluation of fetal well-being, it is important to recognize the difference between the averaged FHR range as determined by periodic auscultation of the fetal heart beat and the instantaneous FHR pattern.[12] The latter is based on a graph which is updated from the R-R intervals of successive FECG's so that each new R-R interval is reflected by a new plotting point. In this way the fine details of FHR changes are presented. If

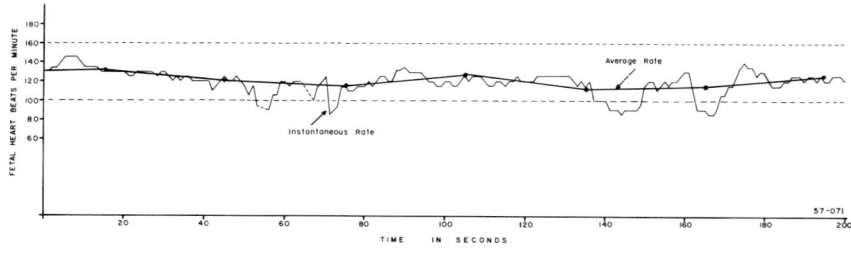

COMPARISON OF INSTANTANEOUS AND AVERAGE FETAL HEART RATES

Figure V-1. The irregular tracing is a graph of the instantaneous FHR while the heavy line is the average of the instantaneous FHR for the preceding thirty seconds. The increased detail of the instantaneous FHR is clearly seen. (From E. H. Hon, *Amer J Obstet Gynec*, 75: 1215-1230, 1958.)

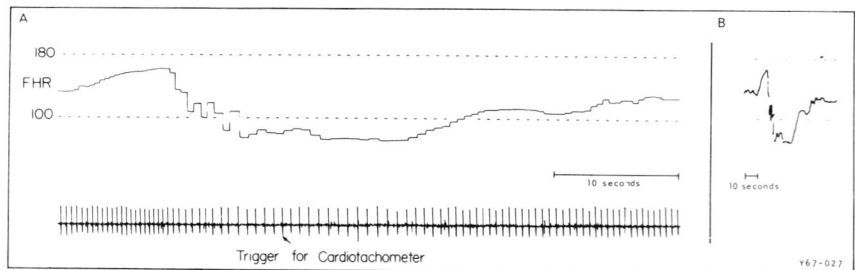

Figure V-2a. The upper trace is the output of a digital cardiotachometer that measures the interval (t) between successive R waves and plots a graph which is proportional to 1/t. The lower trace shows the triggers for the cardiotachometer which are the result of sharply filtering the FECG. *Figure V-2b.* Same data as upper trace of Figure V-2a but plotted on a compressed time scale thereby producing the type of FHR pattern used in this report.

fetal distress

such FHR changes are plotted with a vertical scale of thirty beats per cm and a horizontal scale of 0.33 minutes per cm, FHR patterns of the type shown in this report will result. The scaling factors are important if FHR patterns of one research group are to be compared with another.

The loss of FHR detail as the result of averaging the FHR for thirty seconds is illustrated in Figure 1. The irregular tracing is the instantaneous

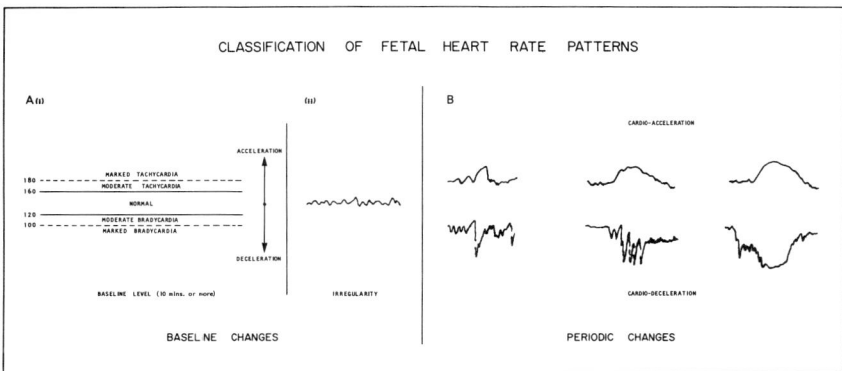

Figure V-3a(i). Method of classifying the FHR between uterine contractions. *Figure V-3a(ii).* Example of baseline FHR irregularity. *Figure V-3b.* Examples of transitory periodic FHR acceleration and deceleration. Since these FHR changes are associated with uterine contractions, they are labeled as being periodic changes because of the quasi-periodic occurrence of uterine contractions.

FHR and the heavy line is the average FHR. The technique for plotting FHR is shown in Figures 2a and 2b. The upper trace of Figure 2a is the output of a digital cardiotachometer that measures the interval (t) between successive R waves and computes a voltage which is proportional to $1/t$. The lower tracing shows the triggers for the cardiotachometer input which have been derived from the R wave of the FECG. The time base of Figure 2b, which is one-tenth that of Figure 2a, is 0.33 minutes per cm.

The current basis for describing FHR patterns is predicated on the concept that each fetus has an FHR baseline level which is peculiar to itself. On this baseline there may be superimposed varying degrees of irregularity illustrated by minor rises and falls in FHR which are the end result of a number

93

prenatal life

of regulating factors impinging on a physiologic control system.[13] In Figure 3A(i) the FHR baseline, which is determined between uterine contractions, has been labeled according to currently accepted clinical criteria: normal 120-160 beats per minute, moderate tachycardia 161-180 beats per minute, marked tachycardia 181 or more beats per minute, moderate bradycardia 100-119 beats per minute, and marked bradycardia ninety-nine or less beats per minute. Baseline irregularity is classified on the basis of the ratio that its peak-to-peak amplitude bears to the level of the baseline FHR rate. The slower and more prolonged (more than ten minutes duration) baseline FHR changes are classified as "tachycardia" or "bradycardia" to distinguish them from transitory rises and falls in FHR which are referred to as "accelerations" or "decelerations" respectively.[13]

FHR acceleration and deceleration patterns which are especially noteworthy, occur in association with uterine contractions and hence are quasi-periodic. FHR acceleration patterns are less clearly defined and less clearly understood than deceleration patterns. The FHR deceleration patterns are well defined and have specific waveforms which are of primary importance in FHR pattern classification. In addition, the onset of each FHR deceleration pattern bears a specific timing relationship to the beginning of the associated contraction. The adjectives "early," "late," or "variable" are used to describe this time relationship and to distinguish periodic FHR deceleration patterns from transitory falls in baseline FHR.[14-17] The early FHR deceleration pattern has its onset early in the contracting phase of the uterus; late FHR deceleration has its onset late in the contracting phase of the uterus, viz. thirty to forty seconds after the beginning of the contraction; variable FHR deceleration has its onset at variable times in the contracting phase of the uterus.

In the diagnosis of fetal distress the type of FHR deceleration pattern present is of utmost importance since it, in many ways, describes the FHR response of the fetus to a repetitive mechanical stimulus which in many instances is a decided stress. The mechanical stress of each uterine contraction may exert pressure on the uterine contents by compression of the fetal body directly, e.g., fetal head, or indirectly alter intervillous space blood flow by impeding venous outflow and arterial inflow. It may also inadvertently cause umbilical cord compression. This concept is illustrated in Figures 4A, 4B, and 4C.

Figure 4A illustrates an FHR deceleration pattern of uniform shape

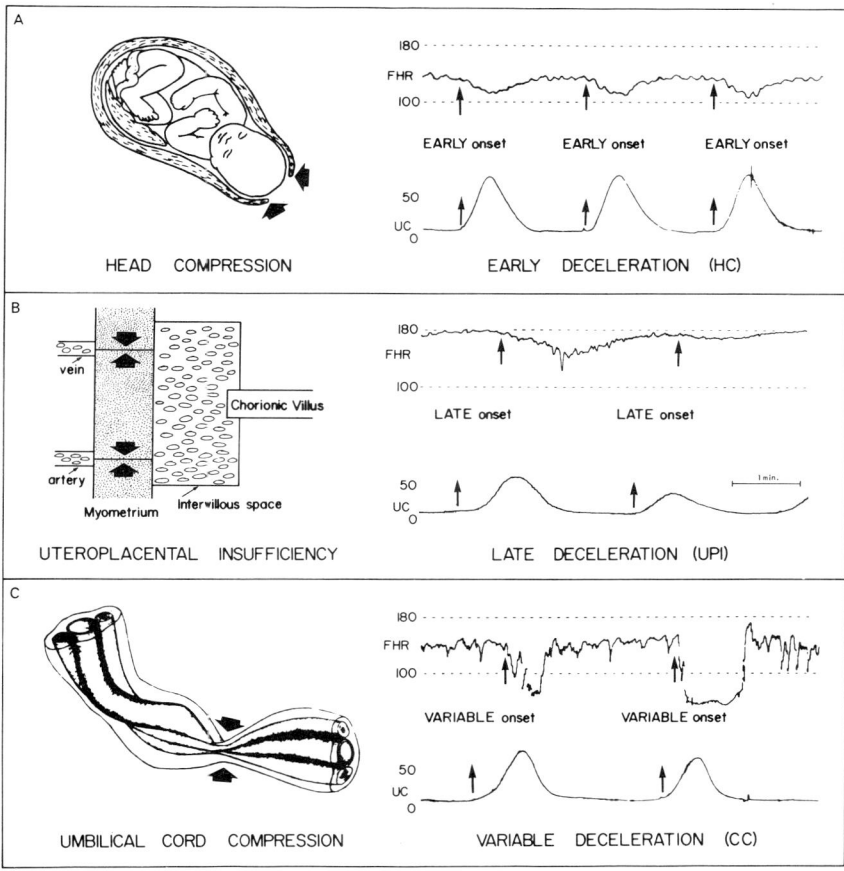

Figure V-4a. Uniform FHR deceleration pattern which reflects the intrauterine pressure curve whose onset begins early in the contracting phase of the uterus. This type of early FHR deceleration is thought to be due to fetal head compression. HC = head compression. UC = uterine contraction. *Figure V-4b.* Uniform FHR pattern which also reflects the shape of the intrauterine pressure curve and which has its onset late in the contracting phase of the uterus. This FHR deceleration pattern is thought to be due to acute uteroplacental insufficiency as the result of interference with intervillous space blood flow by the contracting uterus. UPI = uteroplacental insufficiency. *Figure V-4c.* Variable FHR deceleration pattern whose waveform does not reflect the intrauterine pressure curve and whose onset of deceleration bears a variable time relationship to the beginning of each contraction. This variable FHR deceleration pattern also varies in shape from contraction to contraction and is thought to be due to umbilical cord compression. CC = cord compression.

prenatal life

which reflects the intrauterine pressure curve and whose onset bears a consistent early time relationship to the beginning of each concomitant contraction. This FHR pattern is thought to be due to pressure applied to the fetal head by the dilating cervix.

Figure 4B is an example of another FHR deceleration pattern of uniform shape which also reflects the intrauterine pressure curve, and whose onset bears a consistent late time relationship to the beginning of each contraction. (Contrast this late timing with that of Figure 4A where the consistent time relationship is early.) This late deceleration in FHR pattern is thought to be due to interference with intervillous space blood flow by the contracting uterus and a resultant decrease in maternal-fetal exchange.

Figure 4C is an illustration of the FHR pattern of variable deceleration which is of variable shape and which has an onset that bears a variable time relationship to the beginning of each associated uterine contraction. This type of FHR deceleration pattern does not reflect the intrauterine pressure curve and is thought to be due to umbilical cord compression of varying degrees which is determined by the momentary physical relationship existing between the mother, fetus, and umbilical cord at a specific time.

Some additional characteristics of these three specific FHR deceleration patterns are shown in Figures 5A, 5B, and 5C. Early deceleration is illustrated in Figure 5A. This type of pattern is probably due to the rise in intracranial pressure and alteration in blood flow to the brain stem as pressure is exerted on the soft parts of the fetal skull by the rim of the dilating cervix. This type of pattern is usually seen in primigravid patients or multigravid patients where the cervix is firmer than usual. It has the following characteristics: (1) uniform, specific FHR pattern which reflects the intrauterine pressure curve; (2) onset occurs early in the contracting phase of the uterus; (3) usually does not fall below 100 beats per minute; (4) usually of less than ninety seconds duration; (5) usually associated with a baseline FHR in the normal range; (6) probably due to increased intracranial pressure; (7) not affected by maternal hyperoxia; (8) markedly altered by atropine administration; (9) not associated with fetal biochemistry changes.

The late deceleration pattern due to acute uteroplacental insufficiency is illustrated in Figure 5B. The characteristics peculiar to this pattern are: (1) uniform, specific FHR pattern which reflects the intrauterine pressure curve; (2) onset occurs late in the contracting phase of the uterus; (3) usu-

fetal distress

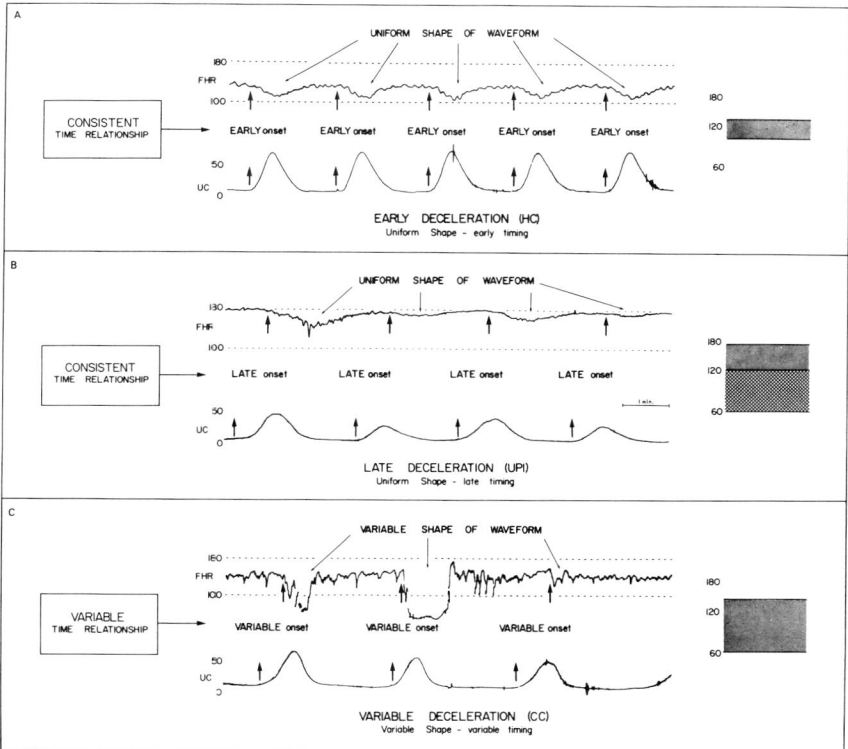

Figure V-5a. FHR pattern of early deceleration showing uniform shape and consistent early time relationship to the beginning of the associated contraction. The cross-hatched area to the right of the figure indicates the usual range in which this type of deceleration is found. *Figure V-5b.* FHR pattern of late deceleration showing uniform shape and consistent late time relationship to the beginning of the associated contraction. This type of FHR deceleration pattern is usually found in the range of 180-120 beats per minute, but if the situation worsens, the rate may drop as low as sixty beats per minute. This is illustrated by the cross-hatched areas on the right of the figure. *Figure V-5c.* FHR pattern of variable deceleration showing variable shape and variable time relationship to the beginning of each associated contraction. The cross-hatched area to the right indicates that this type of FHR pattern is usually in the range of 140-60 beats per minute; if the situation worsens, the duration becomes prolonged.

ally does not fall below 120 beats per minute but may fall to sixty beats per minute or less; (4) usually of less than ninety seconds duration; (5) usually associated with a baseline FHR in the upper, or above normal, range; (6) probably due to uteroplacental insufficiency; (7) usually modified by maternal hyperoxia; (8) partially modified by atropine administration; (9) usually associated with fetal acidosis.

Recent fetal biochemical studies indicate that fetal acidosis is frequently associated with this type of FHR pattern.[18-21] These findings are in agreement with our earlier hypothesis which postulated that impedance of venous outflow from, and arterial inflow to the intervillous space, were responsible for decreased fetal oxygenation which in turn caused this type of late deceleration.[22] The FHR pattern of late deceleration is of clinical importance since it is almost always present with uterine hyperactivity or maternal hypotension. Further, it is observed frequently in toxemia of pregnancy, diabetes mellitus, erythroblastosis fetalis, and in some cases of postmaturity. Clinically, it is deceptive since neither the FHR baseline nor the lowest level of FHR deceleration may go beyond the limits of the normal FHR range of 120-160 beats per minute. This type of deceleration is usually associated with some degree of neonatal depression.[23]

Figure 5C illustrates the FHR pattern of variable deceleration. It has the following characteristics: (1) variable FHR pattern from contraction to contraction; (2) onset occurs at variable times in the contracting phase of the uterus; (3) usually falls below 100 beats per minute, frequently as low as fifty to sixty beats per minute; (4) duration of slowing varies from ten seconds to minutes; (5) usually associated with a baseline FHR in the normal, or low normal range; (6) probably due to umbilical cord occlusion; (7) markedly altered by maternal position change or fetal manipulation; (8) not affected by maternal hyperoxia; (9) markedly altered by atropine administration; (10) not associated with fetal acidosis, unless the FHR changes are frequent and prolonged.

With umbilical cord compression the FHR may be as low as sixty beats per minute. The duration of the deceleration increases and subsequently falls to a lower FHR level if cord compression is prolonged.[24]

Figure 6A is an example of variable deceleration where the FHR falls as low as sixty beats per minute and takes about a minute to return to baseline. Clinically, this type of FHR pattern warrants a diagnosis of "fetal dis-

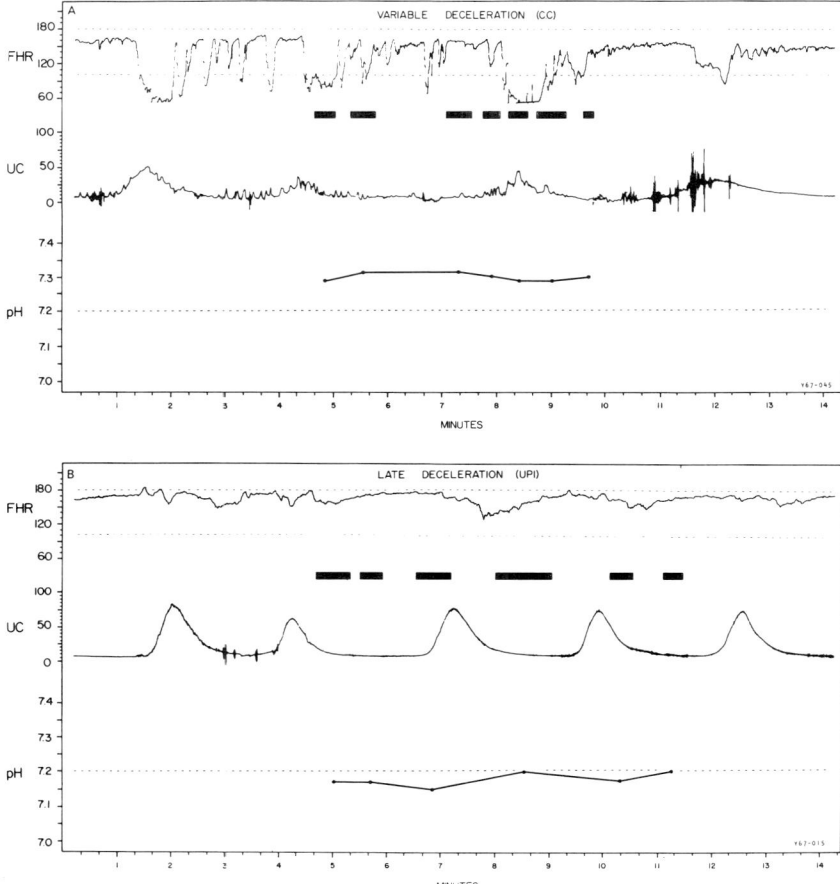

Figure V-6a. Upper trace is an example of the FHR pattern of variable deceleration associated with marked irregularity. This degree of deceleration and irregularity causes undue clinical anxiety and is frequently the cause of unnecessary cesarean section. The pH studies of the fetal scalp blood determined concomitantly are about 7.30 and support the feeling that this degree of short-lived umbilical cord compression is not damaging to the fetus. The shaded areas immediately under the FHR patterns indicate the time of collecting samples from the fetal scalp.
Figure V-6b. An example of an FHR pattern of late deceleration. In this instance, the FHR immediately following contractions is in the present clinically accepted range thereby justifying the conclusion that the fetus was not compromised. However, the pH of the fetal scalp blood determined during this period indicates that it is in the range of 7.15 to 7.20. This, then, is an example of an FHR which is in the normal clinical range, yet the fetus is distressed.

prenatal life

tress;" nevertheless, the concomitant fetal biochemical studies show a pH of about 7.30 in a number of consecutive samples. This finding supports the feeling that this type of FHR pattern is due largely to reflex vagal activity and is of little clinical significance, except that it may serve as a warning sign of probable umbilical cord difficulty and is not within itself an indication for immediate termination of labor.

Figure 6B is an illustration of another FHR pattern which is clinically confusing. This late deceleration pattern was recorded from the fetus of a toxemic patient at about 5 cm dilation. If the FHR is determined stethoscopically at the end of each contraction, as is the usual custom, the FHR will be

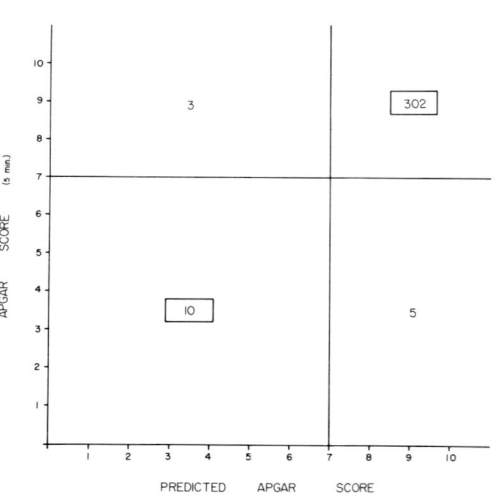

Figure V-7. A diagram showing correlation of Apgar scores at five minutes and the predicted Apgar score based on an assessment of the last twenty minutes of FHR patterns before delivery. Of the 305 patients who had an Apgar score of greater than 7, 302 were correctly predicted. Of the fifteen who had an Apgar score of less than 7, ten were correctly predicted.

found in the clinical normal range. However, the concomitant fetal pH determinations are in the range 7.15 to 7.20 and indicate fetal acidosis. This, then, is an example of a situation where the FHR is clinically normal, yet the fetus is in distress.

Figure 7 shows the results of a study on 320 patients where a correlation between newborn condition immediately after birth and the type of FHR pattern immediately preceding delivery was made. In this study, which was prompted by clinical experience with FHR patterns, the last twenty

fetal distress

minutes of the FHR pattern immediately before delivery were used as a basis for prediction of the Apgar score. On the basis of specific baseline and periodic FHR changes, points are subtracted from a possible Apgar score of 10. The Apgar score at five minutes is used since it correlates better with subsequent infant neurologic deficit than the one minute score.

It is not likely that it will be possible to eliminate entirely the small difference between the predicted and actual Apgar scores, since there are inherent inaccuracies in the Apgar score, and as well FHR changes may not reflect all compromise of fetal well-being. The results obtained with a current prospective study of FHR patterns and prediction of Apgar scores are similar to those obtained with retrospective studies (unpublished data).

Biochemical Studies

The introduction of biochemical analysis of fetal scalp blood by Saling[9] has been a major advance in the study of fetal well-being. Prior to this time, biochemical studies of the fetus were limited to the umbilical cord blood or to the neonate in the immediate newborn period. These earlier studies, together with those of Saling, have provided a broad background for further assessment of FECG, FHR, and neonatal ECG and neonatal heart rate.

Our present combined biochemical and biophysical study of the mother, fetus, and newborn is still in its exploratory phases. However, there appears to be a good correlation between FHR patterns and fetal pH studies if one has a continuous record of the FHR and correlates this with fetal pH changes over a period of twenty to thirty minutes. Even so, there are instances where these techniques for evaluation of fetal well-being are in conflict. In these instances if the Apgar score is a reliable index of fetal conditions, the FHR patterns correlate more closely with the Apgar score than fetal biochemical studies.

Figure 8 is a situation where the FHR pattern is in the normal range and where the fetal pH is about 7.10 for about sixteen minutes. This record was made immediately before delivery. The baby was born with an Apgar score of 8 at one minute and 9 at five minutes. This patient, therefore, illustrates a situation where the pH of the fetal scalp blood is in a definitely pathologic range suggesting severe fetal compromise, yet the infant was born in good condition. A review of the preceding eighteen minutes of the FHR

prenatal life

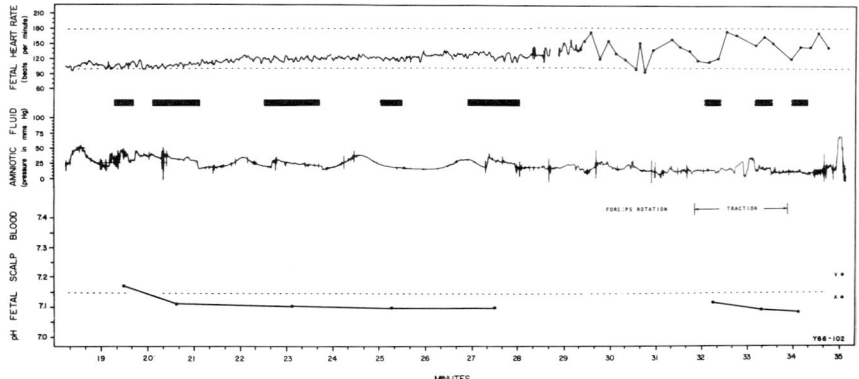

Figure V-8. An example where the FHR pattern is in the normal range and the fetal pH studies determined concomitantly are about 7.10. This record was made immediately before delivery and the baby was born with an Apgar score of 8 at one minute and 9 at five minutes. The apparent reason for the normal FHR pattern and high Apgar scores at birth in the face of concomitant low fetal pH's may be seen in Figure 9.

pattern will provide an explanation for the difference between this normal FHR pattern and the abnormal fetal biochemical findings.

Figure 9 is a record of the FHR pattern and uterine activity immediately preceding the beginning of the tracing shown in Figure 8. Figure 9 shows that there was a period of marked fetal cardiac deceleration of the uteroplacental insufficiency type associated with uterine hyperactivity (resulting

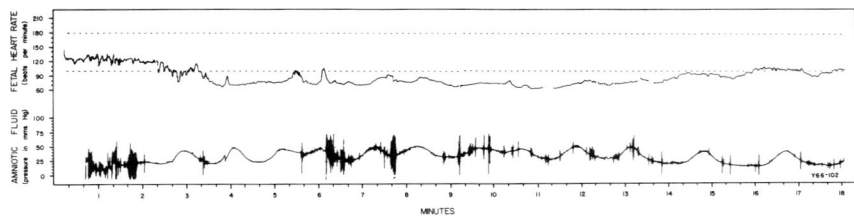

Figure V-9. Marked FHR deceleration pattern indicating acute uteroplacental insufficiency as a result of marked uterine hyperactivity following inadvertent infusion of a large dose of intravenous oxytocin. This abnormal FHR pattern immediately preceded the tracing of Figure V-8 and probably explains the low fetal scalp blood pH shown in that figure.

fetal distress

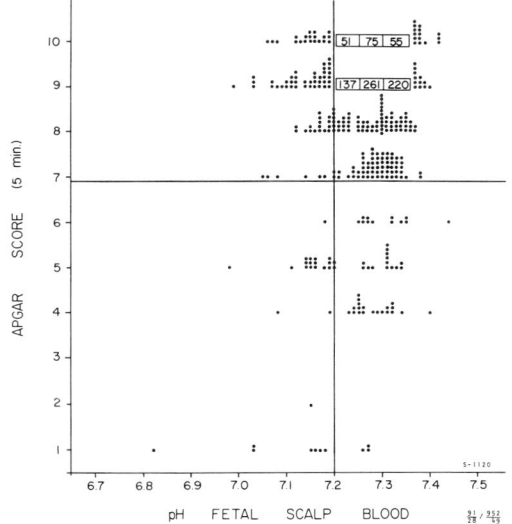

Figure V-10. Correlation of the pH values of 1,120 samples of fetal scalp blood done on 177 patients. Of the 119 determinations where the fetal scalp blood had a pH of less than 7.20, 91 had Apgar score at five minutes of greater than 7. Of the 1,001 samples where the pH was greater than 7.20, forty-nine had Apgar scores at five minutes of less than 7.

from an inadvertent rapid infusion of oxytocin). By the beginning of Figure 8 the FHR pattern had returned to normal but the fetal pH was still recovering from the earlier disturbance. This difference in time constants of the FHR patterns and fetal pH studies has also been observed in other situations. The traces of Figures 8 and 9 emphasize the complementary roles that these two techniques play in fetal assessment. These traces also emphasize the importance of continuous records of FHR in determining the significance of fetal pH changes in a given situation.

prenatal life

Figure 10 further emphasizes the need for evaluation of fetal pH over a period of time rather than using a single sample. This graph correlates 1,120 pH determinations of fetal scalp blood with the Apgar score at five minutes. Of the 119 samples which had a pH of less than 7.20 (and fetal compromise may have been expected), ninety-one were associated with an Apgar score greater than 7. Of the 1,001 samples which had a pH greater than 7.2 (and fetal compromise was not expected), forty-nine were associated with an Apgar score of less than 7. It is clear, then, that consideration of a single pH determination in the evaluation of fetal well-being can be quite misleading. For fetal pH studies to be of maximum value several samples must be taken frequently and compared with biochemical evaluation of a maternal arterial sample drawn at the same time.

Fetal Distress

To date our research group has not accumulated enough experience with fetal biochemical studies to determine the correlation between fetal pH and fetal well-being. However, research studies of FHR patterns for many years and more recent clinical evaluation of FHR obtained from an active obstetrical service indicate that there is a good correlation between FHR patterns, neonatal condition, and later infant followup. Although a great deal of further study needs to be done in this area to obtain a better definition of fetal distress, certain tentative judgments seem warranted since patients have to be managed clinically. The following observations on FHR patterns and their relationship to fetal distress have served as bases for a working clinical hypothesis which is currently being tested: (1) the early deceleration FHR pattern indicating head compression seems to be innocuous; (2) FHR patterns that emphasize the need for close watching of the patient include mild umbilical cord compression and tachycardia of 160 beats per minute or more; (3) signs of fetal distress which are ominous are of two types: (a) variable deceleration FHR patterns of more than one minute in duration, which drop to sixty beats per minute or less and which become progressively worse; (b) late deceleration FHR patterns of any magnitude with or without tachycardia. This situation becomes increasingly critical when tachycardia is present, especially if it is associated with a smooth baseline FHR.

The Treatment of Fetal Distress

(a) Alter the position of the patient. There is no one special position, since repositioning merely represents an attempt to redistribute the mechanical forces of a contraction in such a manner that umbilical cord compression is relieved. In our experience the majority of variable deceleration FHR patterns can be alleviated by repositioning.

The effectiveness of this treatment is illustrated by Figure 11. In the upper trace the patient is on her back, and variable deceleration patterns are clearly seen. In the first portion of the lower trace the patient has been placed on her left side, and the patterns have disappeared. On the right of the lower trace variable deceleration patterns reappear as the patient is returned to a supine position.

Repositioning will also be of benefit if late deceleration patterns due to supine hypotension are present. Movement of the patient to either side will restore normotension. Late deceleration patterns due to uterine hyperactivity are also relieved since uterine contractions are usually decreased by turning the patient from her back to her side.

(b) Administer oxygen at six to seven liters per minute. From our studies variable deceleration patterns, which most frequently are clinically diagnosed as fetal distress, do not seem to be affected by maternal hyperoxia. The late deceleration pattern, on the other hand, which is probably the result of hypoxic depression of the myocardium is uniformly modified by maternal hyperoxia. This is illustrated in Figure 12.

Comment

It is clear that stethoscopic sampling of the FHR is inadequate for the determination of fetal distress, and, although precise criteria for fetal distress have not yet been developed, there is evidence that the application of research techniques to clinical obstetrics will provide information heretofore unavailable.

Continuous monitoring of FHR in high-risk patients has resulted in a smaller percentage of depressed babies and fewer cesarean sections for fetal distress, since every attempt is made to change an abnormal FHR pattern to a normal one. Emphasis must be placed on the FHR pattern

prenatal life

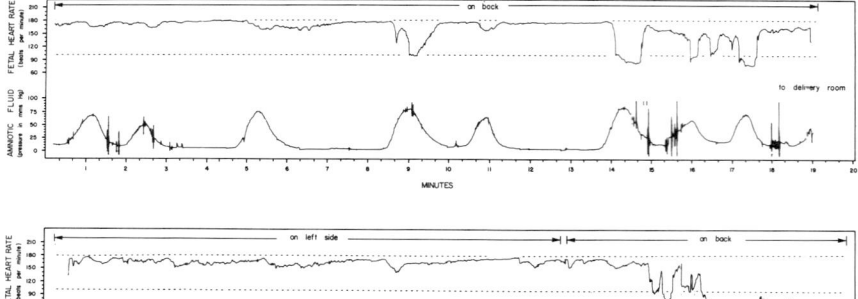

Figure V-11. Upper trace illustrates the variable FHR deceleration pattern of umbilical cord compression with the patient lying on her back. When the patient is turned to her left side, as indicated in the first part of the lower tracing, the variable FHR deceleration pattern disappears. When she is turned on her back again the variable deceleration pattern returns.

rather than the FHR range since late deceleration patterns, which are ominous, frequently fall in the clinically accepted normal FHR range. Cesarean sections for fetal distress are frequently avoided since variable deceleration is responsible for this clinical diagnosis in about 90 percent of our patients. In by far the majority of patients this pattern can be altered by maternal position change.

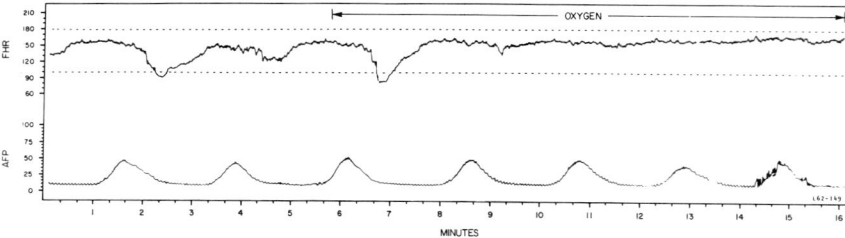

Figure V-12. The early part of the tracing shows the FHR pattern of late deceleration (UPI). Shortly after the administration of 6-7 liters of oxygen to the mother with a tight face mask, there is a marked change in the FHR pattern.

fetal distress

The addition to FHR, recording of an objective measure of uterine activity, obtainable by a transcervical intrauterine catheter, provides a sound basis for evaluating the progress of labor and the tolerance of the fetus to that labor. The use of these combined modalities has taken a lot of guesswork out of the management of the high-risk patient and has given the obstetrician a degree of assurance and control not available with the usual clinical modalities.

Fetal intensive care at the present time should therefore include a technique for producing permanent continuous records of the FHR patterns and uterine contractions, and in situations where doubt still exists, fetal biochemical studies should be used in a complementary fashion.

References

1. Hunter, C. A., Jr., *et al.*: A technique for recording fetal electrocardiogram during labor and delivery. *Obstet Gynec*, 16: 567-570, 1960.
2. Stander, R. W., and Barden, T. P.: Fetal heart rate patterns in normal and abnormal labor. *Nebraska Med J*, 49: 259-264, 1964.
3. Vasicka, A., and Hutchinson, H. T.: Does uterine contractility cause fetal bradycardia? *Obstet Gynec*, 22: 409-418, 1963.
4. Hashimato, K., and Takeda, S.: Fetal electrocardiogram. *Obstet Gynec* (Tokyo), 29: 301-306, 1962.
5. Larks, S. D., and Longo, L. D.: Electrocardiographic studies of the fetal heart during delivery. *Obstet Gynec*, 19: 740-747, 1962.
6. Persianinov, L. S., *et al.*: The dynamics of fetal cardiac activity. *Amer J Obstet Gynec*, 94: 367-377, 1966.
7. Smyth, C. N., and Farrow, J. L.: Present place in obstetrics for foetal phonocardiography and electrocardiography. *Brit Med J*, 2: 1005-1009, 1958.
8. Wood, C., *et al.*: Fetal heart rate and acid base status in the assessment of fetal hypoxia. *Amer J Obstet Gynec*, 98: 62-70, 1967.
9. Saling, E. W.: A new method of safeguarding the life of the foetus during labor. *J Inter Fed Gynaec Obstet*, 3: 100-110, 1965.
10. Beard, R. W., and Morris, E. D.: Foetal and maternal acid-base balance during normal labour. *J Obstet Gynaec Brit Comm*, 72; 496-506, 1965.
11. Kubli, F. W., and Berg, D.: The early diagnosis of foetal distress. *J Obstet Gynaec Brit Comm*, 72: 507-512, 1965.
12. Hon, E. H.: The electronic evaluation of the fetal heart rate: Preliminary report. *Amer J Obstet Gynec*, 75: 1215-1230, 1958.
13. Hon, E. H., and Quilligan, E. J.: The classification of fetal heart rate. II: A revised working classification. *Conn Med*, 31: 779, 1967.

14. Hon, E. H.: Processing of fetal biophysical data. In *Effects of Labour on the Fetus and Newborn*, ed. R. Caldeyro-Barcia, C. Mendez-Bauer, and G. S. Dawes, Oxford, Pergamon Press, 1966.
15. ———: The human fetal circulation in normal labor. In D. E. Cassels, *The Heart and Circulation in the Newborn and Infant*, New York, Grune and Stratton, 1966, pp. 37-52.
16. ———: The detection of fetal distress. International Federation of Gynaecologists and Obstetricians Meeting, Sydney, September, 1967.
17. Hon, E. H., and Quilligan, E. J.: Electronic evaluation of fetal heart rate. IX: Further observations on "pathologic" fetal bradycardia. *Clin Obstet Gynec*, 11: 145, 1968.
18. Quilligan, E. J., Katigbak, E., and Hofschild, J.: Correlation of fetal heart rate patterns and blood gas values. II: Bradycardia. *Amer J Obstet Gynec*, 91: 1123-1132, 1965.
19. Mendez-Bauer, C., et al.: Relationship between blood pH and heart rate in the human during labor. *Amer J Obstet Gynec*, 97: 530-545, 1967.
20. Kubli, F. W.: Personal communication.
21. Hon, E. H.: Unpublished data.
22. ———: Observations on "pathologic" fetal bradycardia. *Amer J Obstet Gynec*, 77: 1084-1099, 1959.
23. Ibarra-Palo, A. A., et al.: Relations between basal fetal heart rate, type II "dips" and Apgar score in the human fetus. In *Effects of Labour on the Fetus and Newborn*.
24. Lee, S. T., and Hon, E. H.: Fetal hemodynamic response to umbilical cord compression. *Obstet Gynec*, 22: 553-562, 1963.

chapter VI

The Environment in which the Fetus Lives: Lessons Learned Since Barcroft

Donald H. Barron, Ph.D.

Inventories of progress are traditionally made at the ends of decades, and two have passed since the English physiologist Sir Joseph Barcroft organized the results of fifteen years of research on the circulation and respiration of the mammalian fetus in a monograph entitled "Researches in Prenatal Life." With that monograph—which has already become a collectors item—he outlined a new field of inquiry and of science, i.e., the physiology of the pregnant uterus and its contents—a field now being actively cultivated by pediatricians and obstetricians, as well as physiologists. This he did by using the facts that he and others had collected to develop concepts and to frame questions for the future, for without concepts or hypotheses there can be no science (there is no science without theory).

Among the concepts that Barcroft advanced one has been of particular interest to obstetricians and pediatricians and of special value as a stimulus to further research: the view that the mammalian fetus develops in an environment in utero in which the availability of oxygen is limited by circumstances over which the fetus can exercise no control. This view holds that the fetus is, so to speak, a victim of its environment—an environment that grows less favorable as gestation advances until at full term, as Barcroft (1947) put it, "the alternatives are escape through birth or death in utero."[4]

My purpose here is to review the principal facts on which this concept

prenatal life

was based, together with some which have been added in the past twenty years, and to take stock of our present concept of the environment in which the fetus lives and the mechanisms by which it is controlled.

As so often happens in research, Barcroft's concept appears to have had its genesis in experiments designed to answer an unrelated question. In the course of his well-known studies on the spleen as a "blood store" Barcroft noticed that during pregnancy (Barcroft and Stevens, 1928) the veins of the uterus and the broad ligament contain relatively large quantities of blood, though their content falls to insignificance soon after birth.[5] To answer the question is the blood in these veins "stored" there during pregnancy, Barcroft and his associates (1933; also 1934) estimated the rate at which it was replaced in the rabbit, i.e., the blood flow through the uterus—and its oxygen consumption at selected stages in gestation.[6,7]

While making these estimates they noted that as gestation advanced after the eighteenth day, the degree to which the blood emerging in the uterine vein was saturated with oxygen fell progressively (Fig. 1). The obvious inference was that after that time the blood flow through the uterus fails to keep pace with the oxygen requirements of the fetuses it contains as they grow and develop. As a consequence the "effective" oxygen pressure in the uteroplacental capillaries drops progressively to lower and lower levels, and Barcroft reasoned that the oxygen tension in the fetal blood leaving the placenta via the umbilical vein must fall concomitantly.

Barcroft was not the first to recognize that the oxygen tension in the fetal blood is low at the end of gestation. Huggett (1927) had shown that

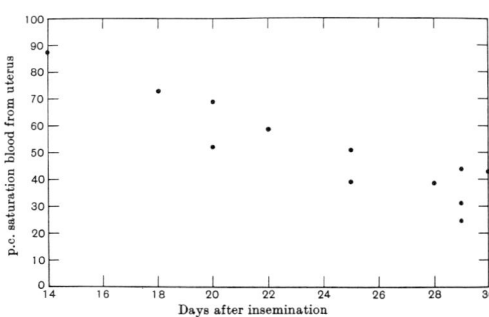

Percentage saturation of blood in uterine veins from the fourteenth to the twenty-eighth day of pregnancy.

Figure VI-1. The change in the oxygen saturation of the blood in the uterine vein of the pregnant rabbit during the course of gestation (Fig. 4, p. 202[6]).

environment of the fetus

to be the case six years earlier in studies on the placenta of the goat.[14] Moreover, Eastman (1930) had made the same observation on human cord blood sampled after vaginal delivery and had suggested that its high oxygen capacity—compared with that of the mother—might be the result of a compensatory response to a low oxygen tension in the source of supply similar to that seen in the acclimatized resident at high altitude.[12] But Barcroft appears to have been the first to suggest that the low oxygen pressure in the fetal blood at the end of gestation is the result of circumstances that change progressively as gestation advances.

It was these observations on the fall in the oxygen saturation in the uterine vein upon which Barcroft (1933) based his generalization that the mammalian fetus develops in an environment in which the oxygen pressure is continually falling as the end of gestation approaches and drew the analogy between the fetus in utero and the mountaineer climbing Mt. Everest.[2] They had, he suggested, a handicap in common: both are living in an environment in which the oxygen tension, over which they have no control, is decreasing toward a level at which survival is impossible. In one case the oxygen is in the air, and in the other it is dissolved in the plasma of the blood in the uteroplacental capillaries, but the circumstances, he reasoned, are essentially the same.

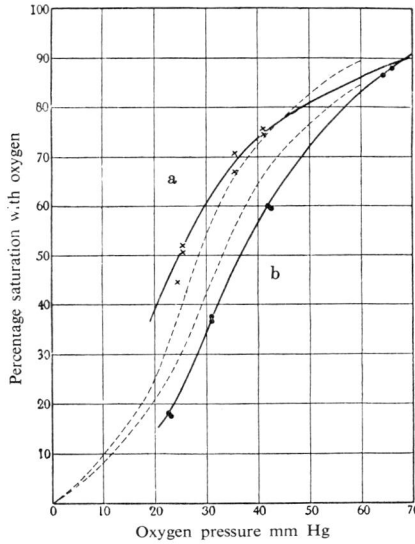

Figure VI-2. A comparison of the oxygen dissociation curves of a) the fetal, and b) the maternal blood of the goat. The dotted lines indicate the limits of curves prepared with blood from nonpregnant does (Barcroft et al.: *Proc Roy Soc* [Biol], *118:* 249, 1935, Fig. 4).

111

prenatal life

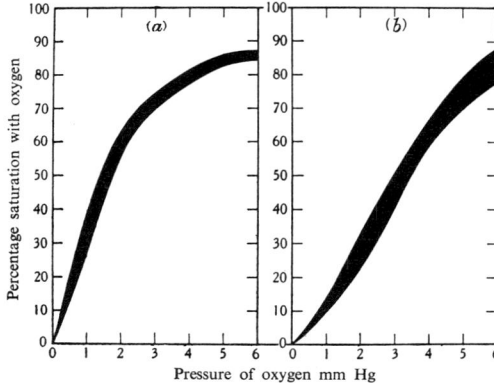

Figure VI-3. The limits of the oxygen dissociation curves of a) the fetal, and b) the maternal hemoglobin of sheep prepared under comparable conditions (Barcroft *et al.*: *Proc Roy Soc* [Biol], *118:* 251, 1935, Fig. 5).

The analogy was a fruitful one, for it prompted the question whether the mechanisms by which the fetus adapts to the decline in the oxygen pressure in its environment are the same as those which operate in the mountaineer and opened the way to further study. But the analogy was not an entirely new one; Eastman (1930) had suggested the parallel earlier.[12] So too had Anselmino and Hoffman (1930) when they advanced their "reacclimatization theory" of the genesis of icterus neonatorum.[1]

To identify the adaptive mechanisms utilized by the fetus Barcroft studied the pregnant sheep and goat. These animals are large enough to provide sufficient fetal blood for study by the methods available, and data were soon forthcoming to indicate that the adaptive mechanisms were indeed very similar to those thought to be characteristic of adaptation to altitude.

First, there was the demonstration that the blood of the fetal goat has

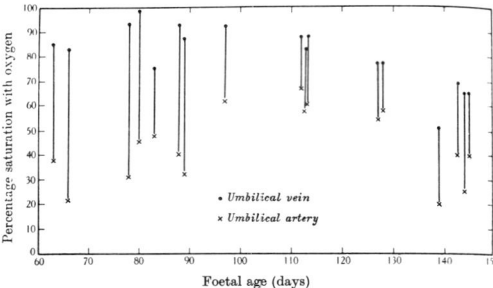

Figure VI-4. The change with advancing gestation in the oxygen saturation of the blood in the umbilical vessels of the fetal sheep as sampled after delivery by caesarean section.[10]

a higher affinity for oxygen than that of the mother (Fig. 2)—that throughout the course of gestation the oxygen dissociation curve of the fetal blood is "to the left" of the maternal (Barcroft et al., 1934; see also McCarthy, 1933) and that this higher affinity is due to the properties of the hemoglobin in the fetal red cells (Fig. 3).[7, 19] This demonstration had an especial importance, for at that time it was thought—erroneously we know now—that a shift in the dissociation curve "to the left" took place in the blood of the mountaineer as a part of the adaptation to high altitude. Other examples of the higher oxygen affinity of the fetal blood relative to the maternal were soon added to the literature to indicate that the phenomenon was a general one. Finally, when Hall and his associates (1936) showed the oxygen affinity of the bloods of birds and mammals indigenous to high altitude to be greater than that in a comparable series which normally live at sea level, the adaptive significance of this feature of the fetal blood appeared to be established beyond question![13]

From his initial observations on the decline in the oxygen saturation in the uterine vein in the rabbit Barcroft had inferred that a concomitant fall in saturation took place in the fetal blood leaving the placenta. And his observations on the oxygen saturation of the bloods entering and leaving, respectively, the fetal side of the placenta of the sheep, after delivery of the umbilical cord by caesarean section, are in full accord with that inference (1940).[10] The oxygen saturation in the vein appeared to rise in the early stages of gestation to a maximum between 100-110 days and thereafter to fall slowly until the end of term (Fig. 4).

Other studies on the umbilical blood of the sheep fetus delivered by caesarean section demonstrated that in these circumstances there is an increase in the oxygen capacity of the fetal blood (Fig. 5) as gestation advances after the 100-110th days that appeared to be inversely related to the oxygen saturation (Barcroft, 1947).[4] Taken together these two lines of evidence appeared to justify the inference that in the sheep the oxygen pressure to which the fetal blood is exposed during its arterialization in the placenta decreases slowly after about the 100-110th day of gestation and that the concomitant increase in the oxygen capacity is a chronic compensatory response to the fall in that pressure as Eastman (1930) had suggested earlier might be the case.[12]

On the basis of this evidence indicating (1) that the oxygen saturation

prenatal life

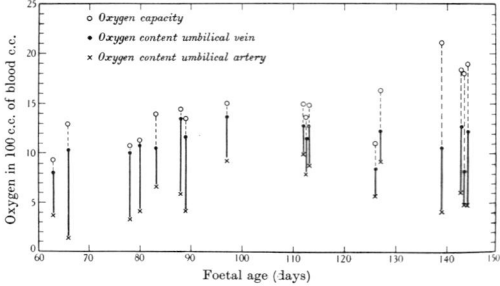

Figure VI-5. The oxygen capacity and content of the blood in the umbilical vessels in the fetal sheep as sampled after delivery by caesarean section.[10]

in the uterine and umbilical veins falls as gestation progresses and (2) that the oxygen capacity of the fetal blood increase concomitantly, it appeared that the placenta as an organ for the supply of oxygen to the fetus failed to keep pace with its demands. Where did it fail? In the rabbit the evidence indicated that the failure lay in the mechanisms that regulate the uteroplacental blood flow, for the uterine blood flow and the weight of the placenta appeared to increase together and to reach a maximum while the fetuses were still quite small.

In the sheep, too, the weight of the placenta—more strictly the weight of the cotyledons—reached a maximum at a stage—about 90-100 days—when the fetus was only a fraction (one-fourth to one-half) of its weight at term

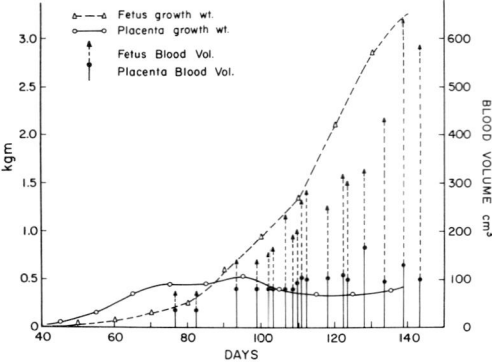

Figure VI-6. The relation in sheep between the volume of blood in the fetal placenta and the weight of the placenta and the fetus during the course of gestation.[4]

environment of the fetus

(Barcroft and Kennedy, 1939; Barcroft, 1947).[9, 4] And estimates of the blood volume in the vascular net on the fetal side of the placenta made after the delivery of the lamb by caesarean section indicated that its capacity reached a maximum at that time—that the functional capacity of the placental portion of the umbilical circuit was related to the weight of the placenta (Barcroft and Kennedy, 1939) and not to the weight of the fetus as an index of its metabolic requirements (Fig. 6).[9] "The vascular bed of the placenta," Barcroft concluded, "is laid down as part of the anatomy of the placenta and not as a part of its physiology" (1947).[4] This apparent failure of the vascular nets of the placenta to grow in proportion to the metabolic needs of the fetus appeared to be responsible for the fall in the oxygen pressure in the umbilical venous blood as gestation advanced.

To put the matter in Barcroft's own words (1944): "The opinion to which I am being forced is that the size of the fetus is limited largely by that of the placenta, more particularly by the placenta as a barrier to oxygen. What it is, however, which limits the size of the placenta is still a question."[3]

Restated in more specific terms this question asks what are the factors that determine the functional capacities of the maternal and fetal vascular beds respectively in the placenta as gestation advances? In search for an answer my colleagues and I found ourselves asking others, e.g., if the fetus carried by a ewe at sea level is, with respect to its oxygen supply living like a man at altitude, how does a fetus manage to survive that is borne by a ewe at high altitude in whose arterial blood the oxygen tension is about half that at sea level? What are the adaptive mechanisms, maternal and fetal, that enable many mammals including man to reproduce at altitudes of 15,000 ft. and higher? Another related question was if the high oxygen affinity of the blood of the adult llama is an adaptation to the low oxygen pressure at altitude is the oxygen affinity of the blood of the fetal llama still further increased?

We went to the Peruvian Andes, to the Instituto de Biologia Andina in Morococha, then under the direction of Professor Alberto Hurtado, to whom we are indebted for his hospitality and cooperation. There we obtained three pregnant llamas and compared the oxygen dissociation curves of the maternal and fetal bloods prepared at the same hydrogen ion concentration (Meschia, et al., 1960).[22] The fetal curve is to the left of the maternal, but the distance between the two is small—much smaller than that between

prenatal life

the bloods of the ewe and her fetus. In utero there may be no difference in the oxygen affinities of the two bloods, because there the fetal blood is a bit more acid than the maternal. This comparison suggests that a significant difference in the oxygen affinities of the maternal and fetal bloods is not essential for the oxygenation of the fetus, even in an animal like the llama whose fetal blood is separated from that of the maternal by a six-layer placenta and in which the effective oxygen pressure in the maternal uteroplacental capillaries appears to be of the order of 25 mm Hg (Meschia, *et al.*, 1960).[22]

The belief that the higher oxygen affinity of the fetal blood is an adaptive response to life in utero appears to be no longer tenable. The characteristics of the hemoglobin of the adult animal are now known to be determined genetically, and it appears that those animals at high altitude whose bloods have a high oxygen affinity selected a suitable environment rather than adapted to an environment in which they found themselves. Moreover, the blood of the immature form of most vertebrate species has a higher affinity for oxygen than that of the adult. It is a feature older in evolutionary history than placentation. It may have enabled the fetus to survive in utero, but it did not appear as an adaptation to life there.

Of even greater interest to us was the observation that the oxygen tension in the umbilical venous blood of the fetal lamb at altitude was not significantly different from that in fetuses studied under similar conditions at sea level (Metcalfe, *et al.*, 1962), despite the fact that the effective oxygen tension in the uteroplacental capillaries was only about 41.9 mm Hg at 15,000 ft., whereas it was 63 mm Hg in our sea level series (Fig. 7).[20] We do not know the rate (ml/kg/min) at which oxygen was actually crossing the placenta at that head of pressure, because we made no estimate of the oxygen consumption of the fetus at altitude, but three lines of evidence taken together suggest that it was not significantly different from the rate at sea level.

First, there was no indication that the fetus at altitude was meeting any measurable fraction of its energy requirements by anerobic metabolism (Huckabee, personal communication). Second, the oxygen consumption of the uterus and its contents appeared to be the same at sea level (10.1) ml/kg/min (Huckabee, *et al.*, 1961) and at altitude (10.3) ml/kg/min (Huckabee, *et al.*, 1959).[15] Third, the growth rate of the fetus carried by the

environment of the fetus

Figure VI-7. A scheme to illustrate the differences in oxygen tension in the maternal and fetal bloods of sheep at sea level and at altitude 14,000 ft. respectively as they move through the placental capillaries.

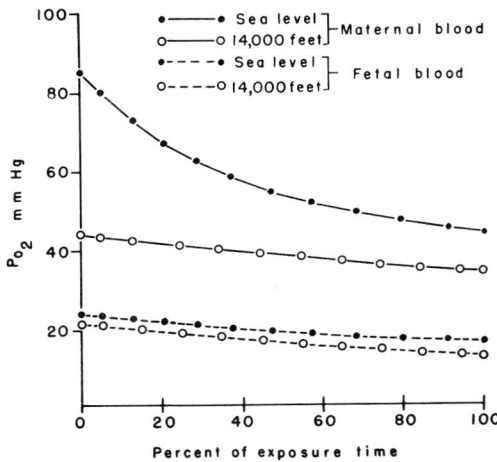

Figure-VI-8. A comparison of the growth rates of lambs carried by ewes maintained at sea level and at altitude respectively (Fig. 2, p. 309[21]).

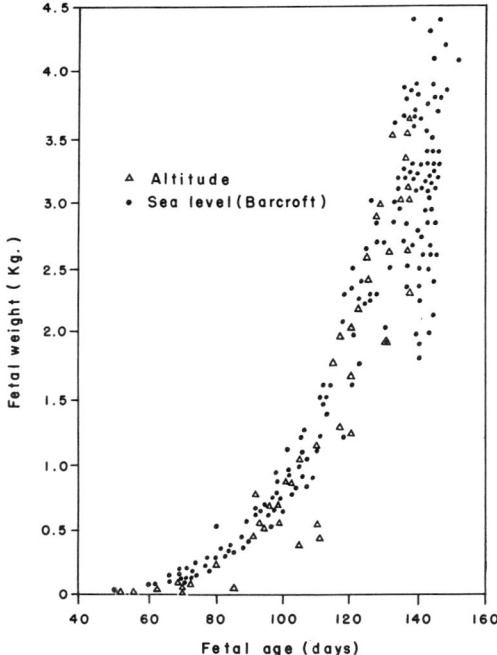

117

prenatal life

ewe at altitude appears to be similar to that of one born at sea level (Metcalfe et al., 1962) (Fig. 8).[20, 21]

On the likely assumption that the oxygen crossing the placenta at sea level and at altitude at comparable stages in gestation is the same (ml/kg/min), it follows that the permeability of the placenta to oxygen must be much greater in ewes at altitude than at sea level. Our results implied that the permeability of the placenta of the ewe to oxygen is not fixed—that it is subject to regulation chronically to the prevailing oxygen pressure in the source from which the fetus obtains its supply, i.e., the maternal blood in much the same way the gill surface of a developing salamander is related to the oxygen tension (Drastich, 1925) in the water in which it is raised or the density of the capillary net on the allantois of the chick is related to the oxygen pressure in the air in which the egg is incubated.[11]

These considerations brought into question the concept that the fetus has no control over its oxygen supply. There was no doubt about the facts upon which the concept was based. Could it be that we had been drawing the wrong inferences from them? Leonardo da Vinci wrote, "Experience is not at fault; it is only our judgment that is in error in promising itself from experience things which are not within her power."

At this point we decided to restudy the functional development of the three components of the placenta involved in the oxygen supply to the fetus —the uteroplacental circulation, the umbilical circulation and the tissues interposed between—over the whole course of gestation in a single species, the sheep, in order to learn how the functional growth of each of these components is regulated.

To have data that are comparable and meaningful for such a study they must be obtained under standard conditions in which the blood flow and oxygen supply of the uterus are not compromised by the demands of other maternal organs in response to stressful circumstances. These conditions we assumed would be met when the ewe was standing quietly, free from apparent stress, in her pen or in a portable stall in the familiar surroundings of the laboratory.

First, we introduced plastic catheters into the uterine veins and the femoral artery and maintained them there chronically to obtain blood samples for the estimation of the arteriovenous differences of oxygen and carbon dioxide differences across the uterus, as well as for the determination of the coefficient of oxygen utilization ($\frac{(a-v)O_2}{aO_2} \times 100$) by its contents at selected stages in gestation in the same animal under standard conditions.

The results of these estimations (Meschia, et al., 1959) demonstrated quite clearly that the blood flow to the uterus is high relative to the oxygen consumption, i.e., the coefficient of oxygen utilization is low prior to about the sixtieth day of gestation (full term about 147 days), and the rate falls relative to that of the oxygen consumption between the sixtieth and eightieth days.[22] Thereafter the coefficient of O_2 utilization appears to be relatively constant until term. That is to say, the blood flow to the uterus tends to increase in proportion to the oxygen consumption of its contents. In ewes bearing singlets the oxygen saturation in the uterine vein often varies within rather narrow limits in the last two months of gestation, although in those carrying twins it may fall gradually toward the end of term.

Our use of indwelling catheters as a means of obtaining blood samples from the uterine vessels opened the way for the estimation of the actual rate at which blood reaches the uterus—again in circumstances in which we believe that rate is not compromised by the demands of other tissues and organs. For these estimations we used the Kety and Schmidt (1948) diffusion-equilibrium method as modified by Huckabee and Walcott (1960) with four amino-antipyrine as the test substance.[18, 16] If repeated estimations were to be made on the same animal, the flow rate was obtained in milliliters per kilogram of tissue perfused—a very useful form for studies of metabolism.

But to estimate the actual volume of blood reaching the uterus we introduced the sampling catheters in the uterine vein and an artery and waited a few days for the ewe to recover before estimating the uterine blood flow in ml/kg/min. The animal was then sacrificed immediately to obtain the combined weight of the tissues perfused, including the uterus, the placenta, and the fetus. With the weight of the perfused tissue and the flow rate per kilogram we calculated the actual flow (Huckabee, et al., 1961).[15]

The averaged results of data obtained in this way, presented in Figure 9, demonstrate clearly that the rate at which blood reaches the uterus of the ewe as gestation advances is related not to the weight of the placenta but to the weight of the fetus.

One further point: although the "uncompromised" rate of uterine

prenatal life

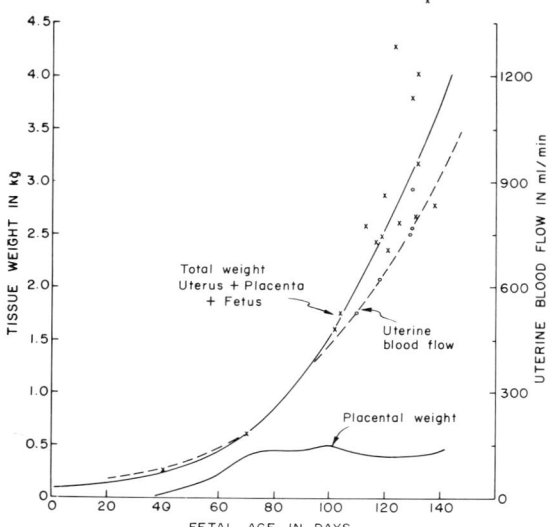

Figure VI-9. A graph to illustrate the relation between uterine blood flow the weight of the uterus and its contents and the weight of the placenta (unpublished data).

blood flow appears to be related to the fetal weight as an index of its demands for oxygen, it should be recognized that the flow can be reduced acutely, as it is in other parts of the spanchnic bed, by mechanisms—neural and hormonal—that are under maternal control, if and when the demands upon her cardiovascular system by other organs in response to stressful circumstances exceed limits as yet undetermined.

Our experience with indwelling catheters in the uteroplacental circulation was so rewarding that it encouraged us to explore the possibility of maintaining them in the umbilical. In the sheep and goat the umbilical vessels are arranged in a fairly regular pattern determined by the distribution of the caruncles—areas of specialization—in the uterine muocsa. As a result it is possible to thread catheters, via tributaries and branches, to cotyledons in the distal part of the uterine horn until the tip is located in the umbilical cord, the common umbilical vein, or the hypogastric artery. The uterine

environment of the fetus

wounds are closed around the catheters, and their peripherial ends are brought out of the abdomen and under the skin to a point on the flank where they are protected by a cover supported by stainless steel pins through the skin.

Through these catheters we have sampled the fetal blood at intervals over a period of weeks without apparent interference with its development, natural birth at full term, or postnatal growth. The results of the analysis of these samples (Meschia, et al., 1965) appear to have provided for the first time data on the oxygen capacity, oxygen and carbon dioxide contents, and the pH of blood in the umbilical circulation when sampled in circumstances shown to be compatible with the continued growth of the fetus—as demonstrated by its subsequent history in utero—and in which the mother and fetus were free from apparent stress.

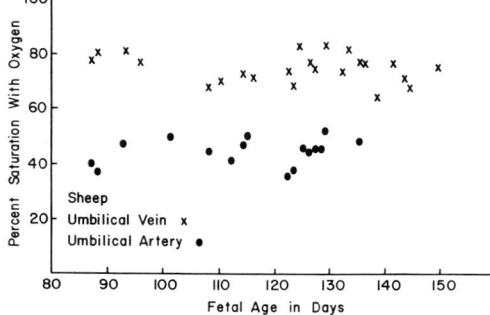

Figure VI-10. The oxygen capacity of the fetal blood sampled via an indwelling catheter is plotted against gestation age in days (Fig. 3, p. 190[25]). Compare with Figure VI-5.

Figure VI-11. The saturation with oxygen of the umbilical arterial and venous bloods respectively of the fetal lamb sampled by indwelling catheters (Fig. 4, p. 192[25]). Compare with Figure VI-4.

121

prenatal life

Contrary to expectation, they provided no evidence that the oxygen supply to the fetus is limited to an increasing degree as gestation advances. The oxygen capacity of the fetal blood remains quite stable in the period between the eighty-seventh and 148th day (Fig. 10). So too do the oxygen saturations and the carbon dioxide contents of the umbilical arterial and venous blood (Fig. 11). Although we have yet to obtain a satisfactory series of estimates of the rate of umbilical blood flow (ml/kg/min) over a period of weeks in a single individual, the relative constancy of the (v-a)O_2 difference across the umbilical circuit observed in an individual fetus for as long as a month and of the oxygen consumption (ml/kg/min) as estimated in a series of fetuses ranging between eighty seven and 148 days gestation age justifies the inference, that, like the flow in uteroplacental circuit, it increases as the fetal weight increases.

The difference between the results of our studies using indwelling catheters and those obtained on fetuses delivered by caesarean section appears to be most easily explained by the absence of stressful circumstances during the sampling of the fetal blood in one series of experiments and their presence in the other. That being the case, the changes in oxygen capacity and content of the umbilical blood after the 100th day of gestation in fetuses delivered by caesarean section can no longer be taken as a response to a chronic fetal hypoxia that increases in severity until birth. To the contrary they suggest that in the acute experiments the circumstances associated with the delivery of the fetus reduced to an increasing degree, as gestation advanced after the 100th day, the oxygen tension to which the fetal blood was exposed during its arterialization.

Be that as it may, our observations on the uteroplacental and umbilical circulations respectively indicate that in the last trimester of gestation the oxygen levels in the capillaries on the two sides of the placenta—and so the transplacental difference in oxygen tension, remain relatively constant. That this is indeed the case is illustrated by values obtained by the analysis of bloods (Meschia, *et al.*, 1965c) drawn via indwelling catheters from the two circulations in the same animal.[26] And the relative constancy of the difference in oxygen pressure on the two sides of the placenta during the last sixty days of gestation—an interval in which the fetal weight increases from about 0.9 to 3.0 kg—suggests that the diffusing capacity of the placenta for oxygen increases as the oxygen requirements of the fetus.

environment of the fetus

Before attempting to determine the diffusing capacity of the placenta for oxygen at selected stages in gestation we tried to estimate the permeability of the placenta to urea, for it seemed technically easier. By infusing urea into the maternal circulation we created a gradient between the uteroplacental and umbilical circuits which changes with time. We followed the changes in urea concentration in the four placental vessels by drawing timed samples via plastic catheters introduced into them as we did in our chronic experiments. The analysis of these samples enabled us to calculate the transplacental difference in urea concentration (mg/ml) for the test period. The quantity of urea crossing the placental barrier (mg/ml) is the product of the v-a difference in urea concentration (mg/ml) across the umbilical circuit and the umbilical blood flow (ml/min).

The results of these experiments (Meschia et al., 1965a) indicate that between the eighty-first and 136th days of gestation the permeability of the placental barrier to urea (Fig. 12) increases markedly and that the increase is linearly related to the increase in the weight of the fetus it serves.[24] Here again it is important to note that the greatest increase in the permeability of the placenta to urea expressed as the diffusing capacity occurs after the placenta reaches its maximum weight.

For reasons I will not detail here, we concluded that the diffusion of urea from the maternal to the fetal blood in our experiments was limited primarily by the trophoblast and that the increase in the diffusing capacity for urea resulted from an increase in the surface area of the membrane rather than a decrease in its thickness. If that is the case, the inference follows that the surface area of the trophoblast increases as does the fetal weight.

As we became more experienced at estimating the diffusing capacity of

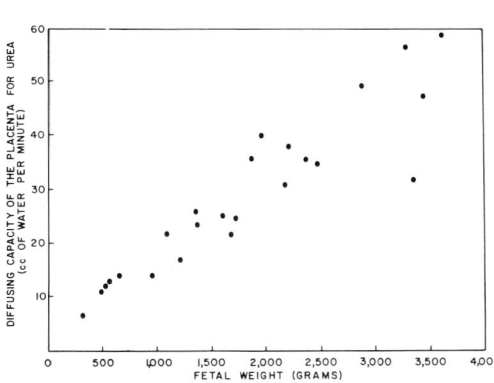

Figure VI-12. The mg of urea that crosses the placenta per minute when the transplacental difference in concentration is mg/ml of water as plotted against fetal weight (Fig. 8, p. 39[24]).

123

prenatal life

the placenta for urea we determined at the same time in some of the later experiments its diffusing capacity for oxygen (Meschia et al., 1965c). This we did by determining the percentage saturation with oxygen and the pH of blood samples from the afferent and efferent vessels of the uteroplacental and umbilical circuits, drawn simultaneously at the beginning and again at the end of the test period.[26] The data referred to appropriate dissociation curves and enabled us to calculate the oxygen tension in the bloods entering and leaving the uteroplacental and umbilical circuits and the difference in the "effective" oxygen pressure on the two sides of the placental barrier.

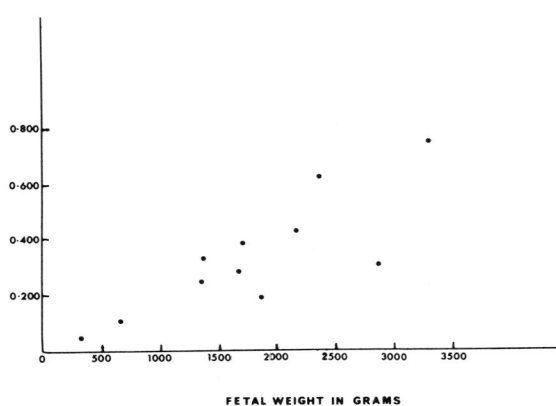

Figure VI-13. The "diffusing capacity" of the placenta for oxygen (ml/min/mm Hg O_2 tension) is plotted against fetal weight (Fig. 4, p. 476[26]).

The oxygen crossing the barrier per unit time was taken as the product of the $(v-a)O_2$ difference across the umbilical circuit and the rate of blood flow through it.

Again our results indicate that there is a linear relation between the permeability of the placenta to oxygen and the weight of the fetus as gestation advances (Fig. 13). They offer no support to the view that the oxygen diffusing capacity of the placenta reaches a maximum around 100-110 days as the observations of Barcroft (Barcroft and Kennedy, 1939) on the volume of blood in the umbilical placental vessels suggest.[9]

In summary, the results of these interrelated studies indicate that (1) in "stress free" circumstances—contrary to our earlier view—the fetus of the ewe lives in a relatively constant environment with respect to the availability of oxygen during the final stages of gestation, and (2) that this constancy is achieved through an increase in the functional capacity of the three units of the placenta directly involved in the oxygen supply of the fetus, which is proportional to the weight of the fetus as an index of its metabolic rate.

The mechanisms by which this adjustment is wrought remain to be described, but there are suggestions that hormones released by the trophoblast—estrogen and progesterone—play an important role in the chronic regulation of the uteroplacental circulation. Interestingly, the first evidence

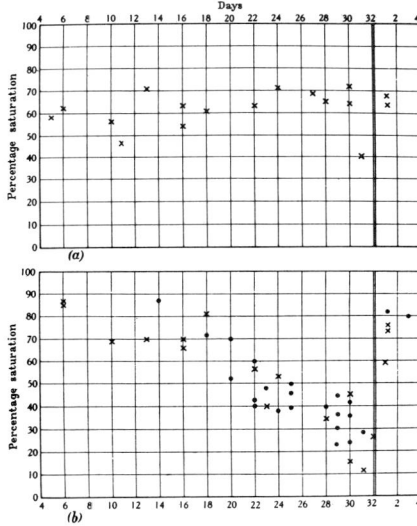

Figure VI-14. A comparison of the oxygen saturation in the venous blood leaving a) the non-pregnant, and b) the pregnant horn of the rabbit as gestation advances. Note equating O_2 sat. with hyperemia in the pregnant horn early in gestation (Fig. 3, p. 219[7]).

in support of this view comes from studies by Barcroft (1934), though he did not draw that inference from his observation.

In the course of repeating his initial experiments on uterine blood flow in the rabbit Barcroft removed one ovary in a series of does before they were bred. As a result, when the does were mated, all the fetuses were located in the horn on the side of the remaining ovary. This made it possible to compare the oxygen saturation of the venous blood emerging from the pregnant and nonpregnant horns. And he observed (Fig. 14) that as early as the sixth day after conception "there is a definite hyperemia on the pregnant side, the emerging blood being perhaps 90 percent saturated." We have confirmed these observations and can add that the hyperemia is not general throughout the pregnant horn. It is confined to those regions occupied by a blastocyst or embryo. As the effect is a local one, it cannot be due to a change in the level of any agent carried to the uterus, and the blood flow is so far in excess of the metabolic requirements of the blastocyst or embryo that it is difficult to conclude that the hyperemia is the result of a response to a local rise in pCO_2 or fall in pH. It appears to be quite akin to the hyperemia seen in the uterus in estrus or after the systemic administration of estrogen.

A similar hyperemia occurs in early pregnancy—prior to the sixtieth day in the sheep—and here again there are strong indications that it is locally induced by an estrogen-like agent released by the blastocyst (Meschia, et al., 1959).[22] There are equally strong hints in the sheep that progesterone released by the placenta plays a role in the chronic regulation of the uteroplacental blood flow in the ewe after the eightieth day of gestation. This is not the time for a review of the details, but clearly this is an area into which research on factors controlling the environment in which the fetus lives is already moving. Another area currently being explored is the nature of maternal mechanisms that control the uteroplacental circulation acutely and the circumstances in which they are activated.

Finally, these concepts and the questions that they provoke are in sharp contrast with those framed by Barcroft with the data available to him twenty years ago. But I am confident that no one would be more excited than he by the change in our point of view as it has been brought about by the lessons from high altitude and from the unstressed preparation. I can only hope our current concepts prove to be as provocative and as fruitful as those which he framed.

References

1. Anselmino, K. J., and Hoffman, F.: Die Urasche des Icterus Neonatorium. *Arch Gynaek, 143:* 477-499, 1930.
2. Barcroft, J.: The conditions of foetal respiration. *Lancet, 2:* 1021, Nov. 4, 1933.
3. ———: Nutritional functions of the placenta. *Proc Nutr Soc, 2:* 14-18, 1944.
4. ———: *Researches in Prenatal Life,* Springfield, Thomas, 1947.
5. Barcroft, J., and Stevens J. G.: The effect of pregnancy and menstruation on the size of the spleen. *J Physiol* (London), *66:* 32-36, 1928.
6. Barcroft, J., Herkel, W., and Hill, S.: The rate of blood flow and gaseous metabolism of the uterus during pregnancy. *J Physiol* (London), *77:* 194-206, 1933.
7. Barcroft, J., et al.: The utilization of oxygen by the uterus in the rabbit. *J Physiol* (London), *83:* 215-221, 1934.

8. ———: Conditions of foetal respiration in the goat. *J Physiol* (London), *83*: 192-214, 1934.
9. Barcroft, J., and Kennedy, J. A.: The distribution of blood between the foetus and the placenta in sheep. *J Physiol* (London), *95*: 173-186, 1939.
10. Barcroft, J., Kennedy, J. A., and Mason, M. F.: Oxygen in the blood of the umbilical vessels of the sheep. *J Physiol* (London), *97*: 347-356, 1940.
11. Drastich, L.: Influence de la pression propre de l'oxygene sur le metabolisme des larves de salamandra maculosa. *Laur Compt rend soc de Biol* (Paris), *92*: 1066, 1925.
12. Eastman, N. J.: Foetal blood studies. I: The oxygen relationships of umbilical cord blood at birth. *Bull Hopkins Hosp*, *47*: 221-230, 1930.
13. Hall, F. G., Dill, D. B., and Gusman, E. S.: Comparative physiology in high altitudes. *J Cell Comp Physiol*, *8*: 301-314, 1936.
14. Huggett, A. St. G.: Foetal blood-gas tensions and gas transfusion through the placenta of the goat. *J Physiol* (London), *62*: 373-384, 1927.
15. Huckabee, W., et al.: Uterine blood flow and metabolism in pregnant sheep at high altitude. *Fed Proc*, *18*: 72, 1959.
16. Huckabee, W., and Walcott, G.: Determination of organ blood flow using 4-amino-antipyrine. *J Appl Physiol*, *15*: 1139, 1960.
17. Huckabee, W., et al.: Blood flow and oxygen consumption of the pregnant uterus. *Amer J Physiol*, *200*: 274-278, 1961.
18. Kety, S. S., and Schmidt, C. F.: The nitrous oxide method for the quantitative determination of cerebral blood flow in man: Theory procedure, and normal values. *J Clin Invest*, *27*: 476-483, 1948.
19. McCarthy, E. F.: A comparison of foetal and maternal haemoglobins in the goat. *J Physiol* (London), *80*: 206-212, 1933.
20. Metcalfe, J., et al.: Observations on the placental exchange of the respiratory gases in pregnant ewes at high altitude. *Quart J Exp Physiol*, *47*: 74-92, 1962.
21. ———: Observations on the growth rates and organ weights of fetal sheep at altitude and sea level. *Quart J Exp Physiol*, *47*: 305-313, 1962.
22. Meschia, G., Wolkoff, A. S., and Barron, D. H.: The oxygen, carbon dioxide and hydrogen-ion concentrations in the arterial and uterine venous bloods in pregnant sheep. *Quart J Exp Physiol*, *44*: 333-342, 1959.
23. ———: Observations of the oxygen supply of the fetal llama. *Quart J Exp Physiol*, *45*: 284-291, 1960.
24. ———: The diffusibility of urea across the sheep placenta in the last two months of gestation. *Quart J Exp Physiol*, *50*: 23-41, 1965a.
25. ———: The hemoglobin, oxygen, carbon dioxide and hydrogen-ion concentrations in the umbilical bloods of sheep and goats as sampled via indwelling plastic catheters. *Quart J Exp Physiol*, *50*: 185-195, 1965b.
26. ———: The diffusibility of oxygen across the sheep placenta. *Quart J Exp Physiol*, *50*: 466-480, 1965c.

chapter VII

Diagnostic and Prognostic Significance of Intrapartum Fetal Tachycardia and Type II Dips

R. Caldeyro-Barcia, M.D., A. A. Ibarra-Polo, M.D., L. Gulin, M.D.,
J. J. Poseiro, M.D., and C. Méndez-Bauer, M.D.

In previous papers[7-10, 29, 30] we have described several patterns in the tracings of FHR which we postulated to be signs of intrapartum fetal distress. These patterns are: (1) high Baseline of the FHR tracing, above 160 beats/min (tachycardia); (2) type II dips, i.e., transient falls of FHR occurring consistently after each uterine contraction (Fig. 1). Type II dips are similar to the "bradycardias of late onset"[19] or the "late decelerations."[23]

The postulate that these two intrapartum FHR changes indicate fetal distress, is based on their association with: (1) hypoxia and acidosis in fetal blood;[10] (2) depression of the newborn; (3) high prevalence of respiratory distress syndrome of the newborn and neurological damage to the child.[10]

We shall make a complete study of the correlations between the condition of the newborn (evaluated by the Apgar score) and the following elements in the FHR tracing recorded during labor: (1) value of the Basal FHR (Baseline); and (2) type II dips. In addition we shall also study the correlations existing between these two elements of the FHR tracing.

This study was supported by USPHS Research Grant HD 00222-06 of the National Institute of Health. Drs. Ibarra-Polo and Gulin were under research fellowships of the Organization of American States.

prenatal life

Correlation between Apgar Score and Basal FHR (Baseline)

This study was made in forty-nine labors in which FHR was graphically recorded with the cardiotachometer[7, 29] from the onset of labor without interruption until delivery. The average value of Basal FHR was calculated for each fetus as follows. The tracings of FHR show at least four different kinds of variations which have been designated as "small rapid oscillations,"[31, 35] "spikes,"[13] "transient ascents"[31] and "dips."[7] The Baseline upon which these variations are superimposed is the Basal FHR[8] (Fig. 1). In the absence of these variations only Basal FHR is recorded.

The Basal FHR is thus measured during the interval between dips, transient ascents and spikes, i.e., when the Baseline is being recorded.

Since the small rapid oscillations are always present in normal tracings, the baseline FHR is estimated as the average value between the peaks and valleys of these oscillations (Fig. 1). In each record hundreds of readings were made, one reading before each uterine contraction. The arithmetic mean of all the values obtained in each record was adopted as the average intrapartum Basal FHR of that fetus during labor.

The condition of the newborn was evaluated by the Apgar score,[1] which was determined in all the infants by the same pediatrician.

The Basal FHR according to the Apgar Score

The forty-nine newborns were classified into two groups according to the first-minute Apgar score[25]—depressed newborns (Apgar score 1-6), and vigorous newborns (Apgar score 7-10). The mean value of the average Basal FHR of all the fetuses was calculated for each of these two groups.

In the group of thirty-four infants which were vigorous at birth (Fig. 2) the mean value of Basal FHR during labor is 143 beats/min (S.D. = 11) (S.E. = 3).[24]

We conclude that the mean value of Basal FHR during labor is significantly higher ($p < 0.001$) in the fetuses which were depressed at birth than in those which were delivered in vigorous condition (Fig. 2).

Figure VII-1. Diagram illustrating some of the variables measured in the tracings of fetal heart rate and of amniotic fluid pressure (uterine contractions) and their chronologic relationships. There is a 40-second lag-time between the peak of uterine contraction and the bottom of the corresponding type II dip. The bottom of the type I dip is recorded almost simultaneously with the peak of the corresponding contraction. Type II dips are usually associated with an abnormal elevation of the Baseline (Basal FHR).[9]

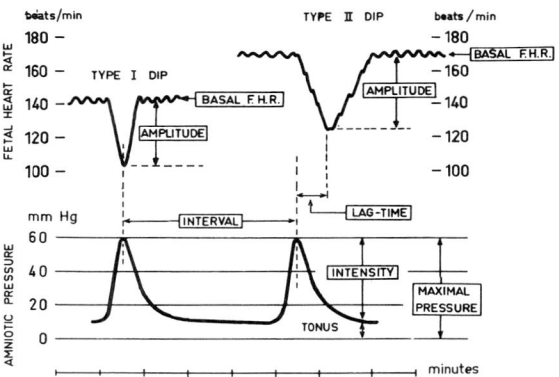

Demarcation Value of Basal FHR between "Vigorous" and "Depressed" Infants

A demarcation value of 155 beats/min has been set for the Basal FHR as follows (Fig. 2):

One S.D. (12 beats/min) is added to the mean (143 beats/min) of the vigorous group (143 + 12 = 155 beats/min).

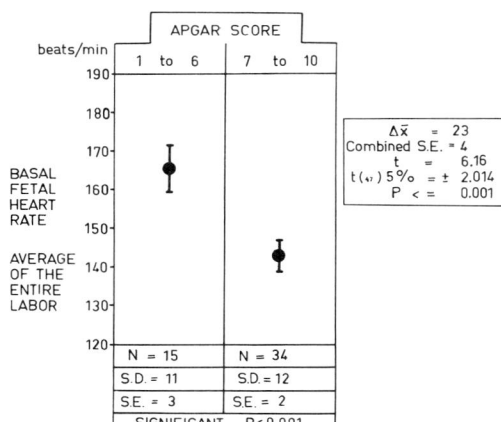

Figure VII-2. Apgar score (first minute) and Basal FHR. The mean value of the "average Basal FHR" of 15 labors which delivered depressed newborns (first minute Apgar score 1 to 6) is 166 beats/min. In the 34 labors which delivered a vigorous infant (first minute Apgar score 7 to 10) the mean is 143 beats/min. Both the mean and its fiducial limits 95% are indicated for each group. The difference between both means is significant (p < 0.001). A summary of statistical analysis is given on the right and also at the bottom of the figure.

prenatal life

One S.D. (11 beats/min) is subtracted from the mean (166 beats/min) of the depressed group (166 − 11 = 155 beats/min).

The Apgar Score according to the Basal FHR

In Figure 3 the average intrapartum Basal FHR of each fetus is plotted against the Apgar score (first minute) of the corresponding newborn.

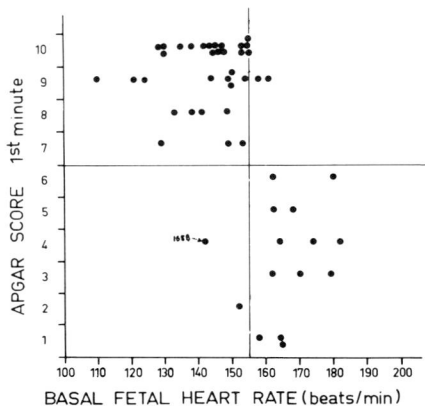

Figure VII-3. Apgar score and Basal FHR. The first minute Apgar score of each of the 49 infants is plotted against the "average Basal FHR" during labor. In 32 out of the 34 vigorous newborns, Basal FHR is below 155 beats/min. In 13 out of the 15 depressed newborns the Basal FHR is above 155 beats/min.

There are thirty-four fetuses in which the average Basal FHR is lower than 155 beats/min (demarcation value). Thirty-two out of these thirty-four fetuses were vigorous at birth; only two were depressed (one of them, record 1658, because the mother had received an excessive dose of Demerol during labor).

There are fifteen fetuses in which the average Basal FHR is higher than 155 beats/min. Thirteen out of these fifteen fetuses were depressed at birth.

The demarcation value of 155 beats/min for Basal FHR proved to be useful for prognostic purposes in forty-five out of the forty-nine infants of the series. When the average intrapartum Basal FHR is above 155 beats/min (tachycardia) the infant was depressed at birth.

They were vigorous at birth when the average intrapartum Basal FHR was lower than 155 beats/min (the lower value has been 115 beats/min).

tachycardia and type II dips

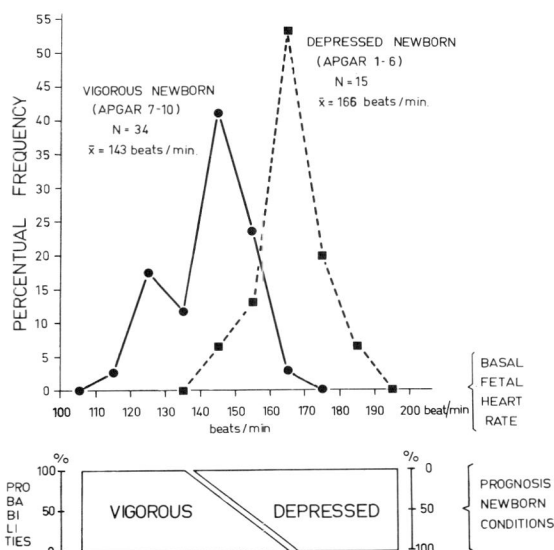

Figure VII-4. Frequency polygons corresponding to the distribution of values of "average Basal FHR" during labor. The polygon on the left side corresponds to the group of 34 vigorous newborns. That on the right to the 15 depressed newborns. The diagram at the bottom represents the probabilities of the infant to be vigorous or depressed at birth, according to the value of "average Basal FHR" during labor.

Prognostic Significance of Basal FHR

Figure 4 shows the frequency distribution of average intrapartum Basal FHR during labor for each of the two groups of infants. The frequency polygon with continuous line corresponds to the group of infants which were vigorous at birth; the polygon with interrupted line to those which were depressed. From Figure 4 it can be deduced that fetuses with average Basal FHR between 120 and 140 beats/min have a very high probability of being vigorous at birth. Fetuses with average Basal FHR between 170 to 190 beats/min have a very high probability of being depressed at birth.[2]

In those fetuses with average intrapartum Basal FHR between 150 and 160 beats/min the probability of being vigorous at birth is only slightly greater than that of being depressed.[2]

The data in the preceding paragraphs and in Figures 3 and 4 indicate that a fetus with an average intrapartum Basal FHR of 130 beats/min has more chances of being delivered in vigorous conditions than a fetus with 155 beats/min.

We conclude that within a range of 120 to 190 beats/min the higher the value of the average Basal FHR (Baseline of FHR tracing) during labor,

133

prenatal life

the greater is the probability that the newborn will be depressed at birth.

These findings agree with the data published[8, 30] that indicate clearly that when the fetus is subjected to a moderate or severe distress the Basal FHR rises.[4, 6, 11, 19, 21, 26, 32]

The rise in Baseline FHR has been interpreted[8, 9] as reflecting an increase in the tonus of the sympathetic system,[8, 9] a feature which, according to our observations, is usually present in fetal distress and is part of the defensive reaction of the fetus.

For practical clinical purposes a tentative limit can be set at 155 beats/min. The rise of intrapartum Basal FHR above this value is considered as a sign of fetal distress. The higher the level of the Baseline and the longer the time during which the Basal FHR remains above 155 beats/min, the worse will be the prognosis for the newborn and for the child.[10, 17, 24]

Results of Other Investigators

Our results agree with those of Fitzgerald and McFarlane,[14] Brady, et al.,[4, 6] who found that tachycardia preceded the delivery of a depressed newborn. Steer[34] and Walker[36] found that the perinatal mortality was higher in fetuses that had intrapartum tachycardia.

Our results disagree with those of Lund,[27, 28] Fenton and Steer,[12] and Ginsburg, et al.,[15, 16] who reported no association between intrapartum tachycardia and the condition of the newborn.

The disagreement among various investigators may be accounted for by the differences in the methods employed for evaluating the FHR and the condition of the newborn.

Post-atropine Basal FHR and Apgar Score

In previous publications[9] we have postulated that the more severe the fetal distress, the higher should be the level reached by FHR after injection of atropine to the fetus.

The objective evidence enabling such a statement is presented in Figure 5 in which the Apgar score (first minute) is plotted against the value reached by FHR after the injection of atropine. In the eight infants which were delivered in good conditions, the post-atropine FHR was lower than

176 beats/min. In the seven infants which were depressed at birth, the post-atropine FHR was higher than 177 beats/min. In one fetus, which died in utero, post-atropine FHR reached 220 beats/min.

An inverse linear correlation is found between Apgar score and post-atropine FHR.[33]

Correlation between Intrapartum Type II Dips in Fetal Heart Rate and the Apgar Score

A type II dip is a transient fall of FHR produced by one uterine contraction and occurring thirty to fifty seconds after the peak of the contraction (Fig. 1). Type II dips have been extensively described in previous papers.[7-10, 29, 30]

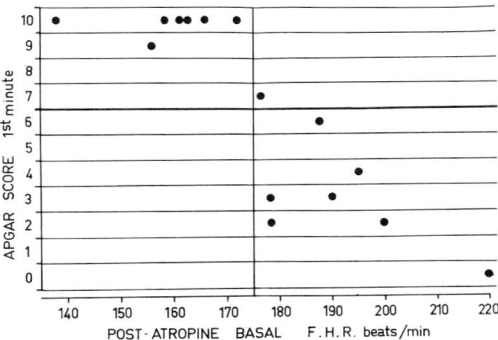

Figure VII-5. Apgar score and post-atropine Basal FHR. The first minute Apgar score is plotted against the value reached by FHR after atropine was injected into the fetal buttock. Atropine causes an average rise of 20 beats/min in FHR. In the 7 depressed infants post-atropine Basal FHR is higher than 175 beats/min. In 7 out of the 8 vigorous newborns, post-atropine Basal FHR is lower than 175 lat/min.

The present study was made in forty-eight cases in which FHR was recorded electronically without interruption from the onset of labor until delivery. The infants were divided into two groups (vigorous and depressed) according to their condition at birth as evaluated by the first-minute Apgar score (depressed, Apgar 1-6; vigorous, Apgar 7-10).[25]

In this paper we shall study the correlation between the Apgar score of the newborn and the following quantitative characteristics of type II dips:[17, 24] (1) total number of type II dips recorded during labor; (2) percentage of uterine contractions which produced type II dips during labor; (3) sum of amplitudes of all the type II dips recorded during labor; and (4) mean amplitude of type II dips recorded during labor.

135

prenatal life

Apgar Score and "Total Number" of Type II Dips Recorded During Labor

In the group of thirty-four vigorous newborns (Fig. 6) the mean values of the total number of type II dips recorded during labor is eleven (S.D. = 17; S.E. = 3).

In the group of fourteen depressed newborns the mean value is ninety-one type II dips (S.D. = 91; S.E. = 24). The statistical analysis ("*t*" test)

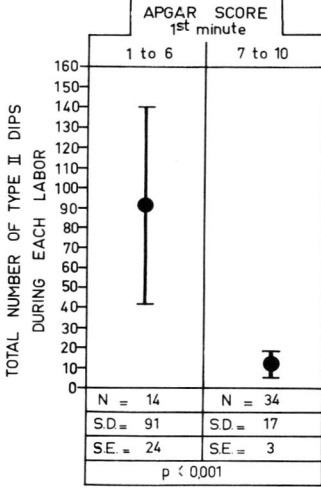

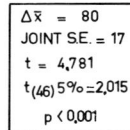

Figure VII-6. The mean value of the "total number" of type II dips recorded during labor in the group of 14 depressed newborns (Apgar score 1 to 6) is 91. In the group of 34 labors which delivered vigorous infants the mean value is 11. Both the mean and its fiducial limits (95%) are indicated for each group. The difference between both means is significant (p < 0.001). A summary of statistical data is given on the right and also at the bottom of the figure.

shows that the mean value of the total number of type II dips recorded during labor is significantly higher ($p < 0.001$) in the group of cases with depressed newborns than in the group with vigorous newborns (Fig. 6).

A tentative demarcation value between the two groups has been temporarily set at twenty type II dips during labor. In Figure 7 the Apgar score of each newborn is plotted against the total number of type II dips recorded during labor.

In thirty out of the thirty-four vigorous infants the number of type II dips was less than twenty. (In five of these thirty infants not a single type II dip was recorded during the whole duration of labor). Only in four vigorous

tachycardia and type II dips

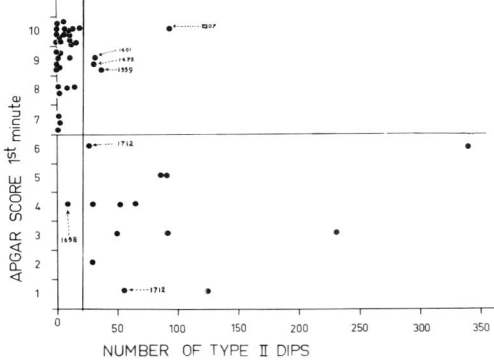

Figure VII-7. The Apgar score is plotted against the "total number" of type II dips recorded during labor. In 30 out of 34 vigorous infants less than 20 dips had been produced. In 13 out of 14 depressed infants more than 20 dips occurred during labor.

infants (Nos. 1401, 1559, 1475, 1207) was the total number of type II dips greater than twenty. These four cases will be analyzed in the discussion.

In fourteen out of the fifteen patients of whom depressed newborns were delivered, the total number of type II dips recorded during labor was higher than twenty. In the remaining case (No. 1658) the intrapartum number of type II dips was lower than twenty; the newborn was depressed perhaps because of the excessive dose of Demerol received by the mother during labor.

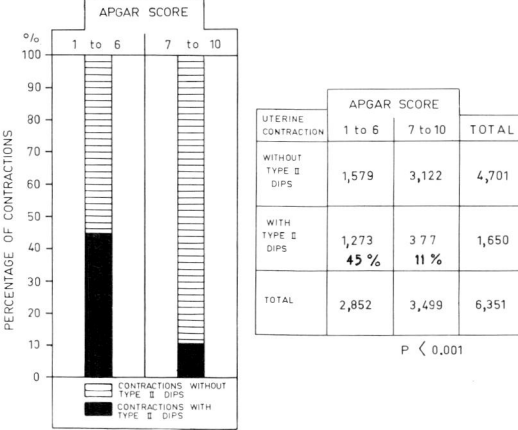

Figure VII-8. In the 14 labors which delivered depressed newborns, 45% of the uterine contractions caused type II dips. In the 39 labors which delivered vigorous newborns, only 10% of the contractions caused type II dips. According to the Chi square test $p < 0.001$. The total number of contractions analyzed is 6,351.

UTERINE CONTRACTION	APGAR SCORE		TOTAL
	1 to 6	7 to 10	
WITHOUT TYPE II DIPS	1,579	3,122	4,701
WITH TYPE II DIPS	1,273 *45%*	377 *11%*	1,650
TOTAL	2,852	3,499	6,351

$p < 0.001$

137

prenatal life

We conclude that the majority of the depressed newborns had more than twenty type II dips during labor, whereas the reverse is true for the vigorous newborns.

Apgar Score and "Percentage of Uterine Contractions" which Produced Type II Dips During Labor

This problem was approached in two different ways. In the first method a pool was made with the 3,499 uterine contractions recorded in the thirty-four labors in which the newborns were delivered vigorous (Fig. 8). It was found that only 377 of the contractions (i.e., 11 percent) caused type II dips, whereas the remaining 3,122 contractions had no such effect.

In the fourteen labors with *depressed* newborns (Apgar 1-6) 1,273 out of 2,852 contractions (45 percent) caused type II dips. The statistical analysis (chi square) showed lack of independence between Apgar score and percentage of contractions causing type II dips ($p < 0.001$). The statistical analysis shows that the percentage of uterine contractions causing type II dips is significantly higher in labors with depressed newborns than in labors with vigorous infants.

In the second approach the Apgar score of each baby was plotted against the percentage of uterine contractions which produced type II dips in that particular labor (Fig. 9). In thirty-three out of the thirty-four cases with vigorous newborns the percentage of contractions causing type II dips was lower than thirty-five percent. The only exception was case 1207, which had Apgar 10 despite the fact that 50 percent of contractions caused type II dips. Case 1207 is analyzed later in the discussion.

In eleven out of the fourteen cases with depressed newborns the percentage of contractions causing type II dips is higher than 35 percent (the highest percentage observed is 80 percent). There are three exceptions in which the percentage is lower than 35 percent. These are cases 1712, 1715, and 1658, which will be analyzed later in the discussion.

The newborns were depressed in eleven out of the twelve cases in which more than 35 percent of the uterine contractions caused type II dips.

This analysis shows that the infants have a high probability of being depressed at birth when more than 35 percent of the contractions cause type II dips during labor.

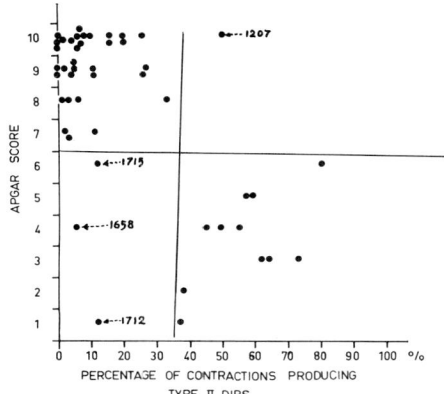

Figure VII-9. In 33 out of the 34 cases with vigorous newborns, the percentage of contractions which produced type II dips during labor is lower than 35%. In 11 out of the 14 cases with depressed newborns that percentage is higher than 35%.

Apgar Score and Sum of Amplitudes of Type II Dips Recorded During Labor

The amplitude of each type II dip is the difference in FHR between the Basal FHR preceding the dip and the FHR at the bottom of the dip (Fig. 1). For example, the type II dip illustrated in Figure I has an amplitude of (170 beats/min—125 beats/min = 45 beats/min or "amplitude units"). To avoid confusion, in the future we shall indicate the amplitude of the dips in amplitude units instead of beats/min.

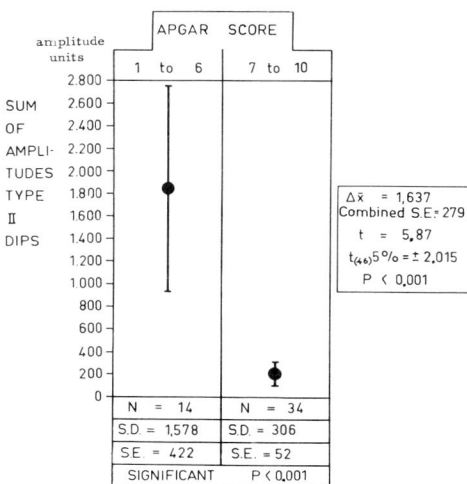

Figure VII-10. The "sum of amplitudes" of type II dips in the 14 infants which were depressed at birth has a mean value of 1,842 "amplitude units." The mean value for the group of 34 vigorous infants is 205 "amplitude units." Both the mean and its fiducial limits 95% are indicated for each group.

The difference between both means is significant (p < 0.001). Statistical analysis is summarized on the right and also at the bottom of the figure.

prenatal life

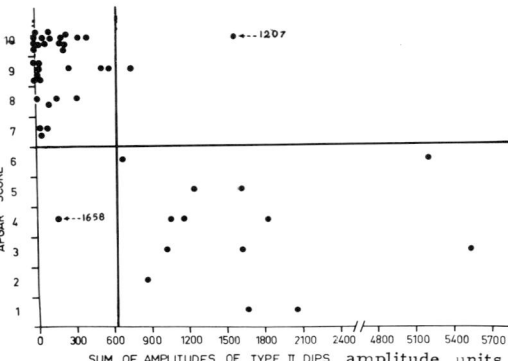

Figure VII-11. In 32 out of the 34 infants which were vigorous at birth, the "sum of amplitudes" of type II dips during labor is smaller than 600 "amplitude units." In 13 out of the 14 depressed newborns, the "sum of amplitudes" is greater than 600 "amplitude units."

In each FHR tracing we measure the amplitude of all the type II dips recorded during a given labor. The amplitudes of all the dips are added, and the resulting number is known as sum of amplitudes of type II dips, which is also expressed in amplitude units. The sum of amplitudes depends on the total number of type II dips recorded in that given labor, as well as on each individual amplitude.

Mean values for the sum of amplitudes were calculated for each of the two groups of infants, vigorous and depressed (Fig. 10). In the group of thirty-four vigorous infants the mean value of the sum of amplitudes was 205 amplitude units (S.E. = 52).

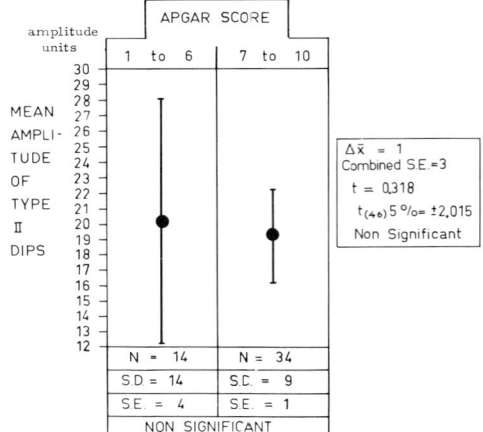

Figure VII-12. There is no significant difference in the mean values of the "mean amplitude" of type II dips between the group with vigorous and with depressed newborns. In both the "mean amplitude" is about 20 "amplitude units."

140

In the group of fourteen depressed newborns the mean value of the sum of amplitudes was 1842 amplitude units (S.E. = 422).

The statistical analysis ("t" test) shows that the mean value of the sum of amplitudes of type II dips recorded during labor is significantly higher ($p < 0.001$) in the group with depressed infants than in the group with vigorous newborns.

A tentative demarcation value between the two groups (depressed and vigorous) has been temporarily set at 600 amplitude units.

In Figure 11 the sum of amplitudes of type II dips recorded during each labor has been plotted against the corresponding Apgar score of the newborn. In thirty-two out of the thirty-four vigorous infants, the sum of amplitudes was lower than 600 amplitude units. Only two labors corresponding to vigorous newborns had a sum of amplitudes greater than 600 amplitude units. These are cases 1207 and 1475 that are analyzed later in the discussion.

In thirteen of the fourteen depressed newborns the sum of amplitudes was greater than 600 amplitude units (the highest value recorded was 5500 amplitude units). There was only one depressed newborn with a sum of amplitudes smaller than 600 amplitude units (case 1658, which received an excessive dose of Demerol).

We conclude that the infant has a very high probability of being born depressed when the sum of amplitudes of type II dips recorded during labor is greater than 600 amplitude units.

Apgar Score and Mean Amplitude of Type II Dips

The mean amplitude of type II dips is calculated for each labor by dividing the sum of amplitudes by the total number of type II dips. Figure 12 shows that there is no difference between vigorous and depressed infants in the mean amplitude of type II dips. In both groups the mean amplitude is about 20 amplitude units.

In Figure 13 the mean amplitude of type II dips of each labor is plotted against the corresponding Apgar score of the newborn.

No difference is seen between the groups with Apgar score 7-10 and those with Apgar score 1-6. The only case which clearly diverges from the average is a fetus with a mean amplitude of 65 amplitude units and which at birth had an Apgar score of 4.

prenatal life

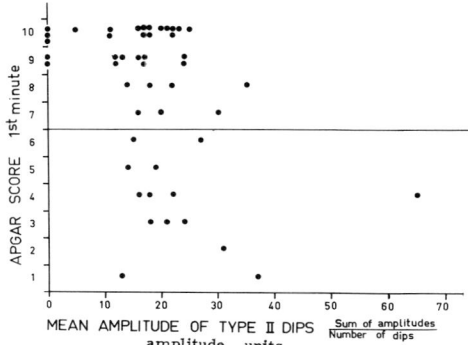

Figure VII-13. No differences are found in the "mean amplitude" of type II dips between the cases with depressed and with vigorous newborns. Five out of the 34 vigorous infants had not one single type II dip during labor (mean amplitude equal to 0). One depressed infant had a very large mean amplitude (65 "amplitude units").

The results illustrated in Figures 12 and 13 show that no differences have been found in the mean amplitude of type II dips between labors delivering depressed newborns and those with vigorous newborns.

These findings indicate that when type II dips are present, their mean amplitude (20 amplitude units) is about the same in normal and distressed fetuses.

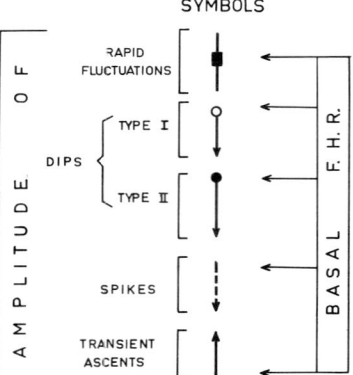

Figure VII-14. Meaning of the symbols employed to illustrate the values and variations of the FHR in the perinatal charts of Figures 15 and 16.[30]

Discussion
Results of Other Investigators

Although the type II dip is a parameter not frequently analyzed as such in the literature, results obtained by other authors may be compared with ours.

Hon[18-21] described transient falls of FHR which occur after uterine contractions ("late onset bradycardias"). These falls of FHR may be considered as equivalent to our type II dips. Although Hon has not yet correlated the presence of late onset bradycardias with the condition of the newborn, the analysis of his published material shows that his late onset bradycardias were present usually in labors that delivered a low Apgar score infant.

Hon[22] developed a system for the prediction of the Apgar score based on the last twenty minutes of the FHR pattern obtained immediately before delivery. Points were subtracted from a possible Apgar score of 10 on the basis of Baseline FHR and on the effects of uterine contractions on FHR. The criteria accepted by Hon are very similar to our own. Of the 305 infants who had an Apgar score of 7 or more, 302 were predicted correctly, and of the fifteen who had a score of 6 or less, ten were predicted correctly.

Brady, et al.[4,5] and Ginsburg, et al.[16] also found transient falls of FHR similar to our type II dips during labors that delivered depressed newborns. Berendes[3] reported that the magnitude and number of consecutive reductions in FHR is greater in cases with depressed newborns or neonatal death. If we assume that these variations in FHR correspond to type II dips, these authors' results would be in agreement with the findings reported here.

Analysis of Divergent Cases from the General Correlations between Intrapartum FHR and Apgar Score

Four infants were delivered in good condition (Apgar score 9 or 10), despite having had more than twenty type II dips during labor. Three of these cases (records Nos. 1401, 1475, 1559) had between twenty-five and thirty-five type II dips (Fig. 7). The fourth case (record No. 1207) had ninety-four type II dips (Fig. 7), 50 percent of uterine contractions caused type II dips (Fig. 9) and the sum of amplitudes of type II dips was 1,500 amplitude units (Fig. 11). All these values of record No. 1207 fell above the tentative upper limits set for the group of vigorous newborns.

prenatal life

In all these four cases the type II dips were produced only in early stages of labor, several hours before delivery, when a cause of fetal distress was acting for a limited period of time. For example, in case 1207 the cause of fetal distress was a marked fall of maternal arterial pressure, which lasted for three hours. When arterial pressure gradually recovered its initial value (see Perinatal Chart of case 1207 in Fig. 15), fetal distress was corrected, and type II dips disappeared from the FHR tracing. No type II dips were recorded in case 1207 during the last four hours of labor during which the FHR tracing had a normal pattern (Fig. 15). The fetus had been restored to normal conditions before delivery, and the infant was vigorous at birth.

The individual analysis of these four cases indicates that, in addition to an evaluation of type II dips during the entire duration of labor, a separate evaluation should be made of the events occurring during the last two hours preceding delivery, since these are more closely related to the condition of the newborn.

In two depressed newborns (cases 1712 and 1715) the type II dips appeared only in the last two hours of labor. (The last part of the perinatal Chart of case 1712 is shown in Fig. 16.)

In these two fetuses more than twenty type II dips were recorded during labor (Fig. 7) and the sum of amplitudes was greater than 600 amplitude units (Fig. 11). These two variables thus fell within the expected range for depressed newborns. However, the percentage of contractions which produced type II dips is lower than 35 percent in both cases (Fig. 9), because during the early stages of labor several hundreds of contractions were recorded that did not produce type II dips (Fig. 16).

The analysis of cases 1712 and 1715 again emphasizes the importance of making a separate study of the last two hours of labor, since fetal distress developing during this period will be sufficient to cause a depression of the newborn, despite the absence of distress during earlier stages of labor.

Case 1658 is an example of pure pharmacologic depression of the newborn. The intrapartum FHR record showed only nine type II dips (Fig. 7), the sum of amplitudes was 166 amplitude units (Fig. 11), and only 5 percent of the contractions caused type II dips (Fig. 9).

In addition, the average intrapartum Basal FHR of this record (142 beats/min) was normal (Fig. 3). Briefly, fetus No. 1658 had no intrapartum signs in FHR which would indicate fetal distress. However, during labor the

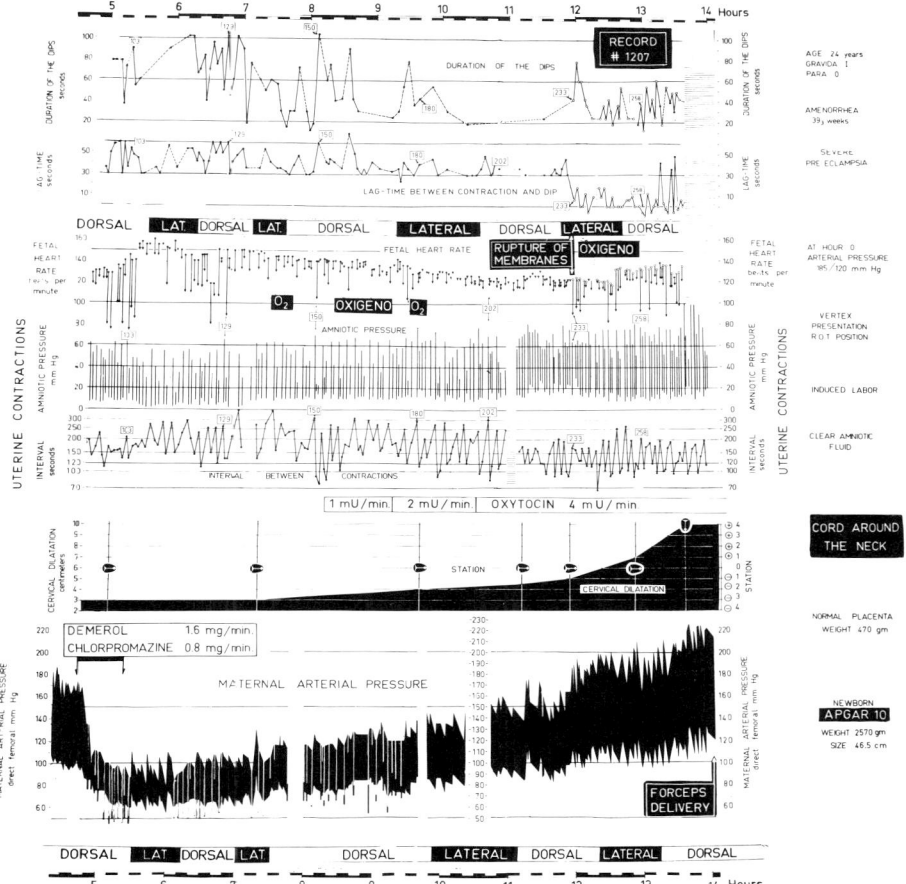

Figure VII-15. Perinatal chart of labor 1207. Severe preeclampsia with arterial hypertension. The i/v infusion of Demerol and Chlorpromazine caused a marked fall of arterial pressure and appearance of "Poseiro effects." Clear signs of fetal distress appeared in the FHR tracings, namely type II dips of large amplitude and rise in the Baseline (hour 5:00 to 8:00).

Two hours after the end of the infusion of Chlorpromazine, arterial pressure started to rise slowly and progressively until fully recovering its initial high values. Concomitantly the signs of fetal distress disappeared from the FHR tracing, the amplitude of type II dips diminished until it became negligible, and the Baseline returned to its original normal levels.

After rupture of membranes each uterine contraction caused a type I dip, which has no ominous connotation. During the last 3 hours of labor preceding delivery FHR tracing shows no sign of fetal distress. A vigorous newborn (Apgar 10) was delivered at hour 14:00, despite the episode of severe fetal distress which occurred between hour 5:00 and 8:00.

(After J. Bieniarz, R. Fernández-Sepúlveda, and R. Caldeyro-Barcia: Effects of maternal hypotension on the human fetus, II: Fetal heart rate in labors associated with cord around the neck and toxemia. *Amer J Obstet Gynec, 92:* 832-849, 1965.)

145

prenatal life

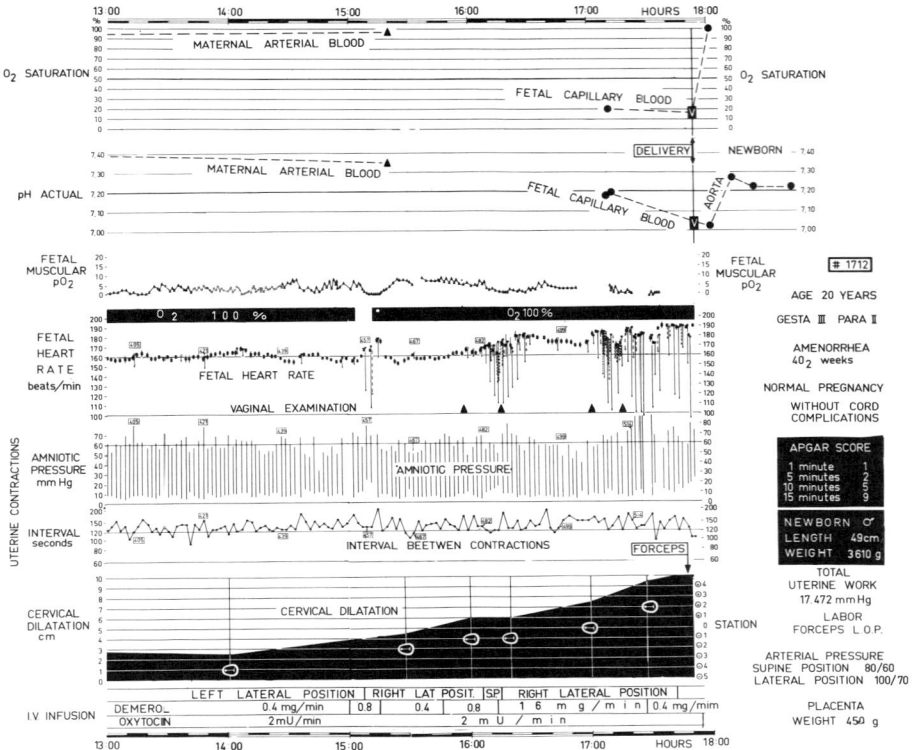

Figure VII-16. Perinatal chart (part III) of labor 1712. Until hour 16:00 there is fetal tachycardia without type II dips (moderate fetal distress). From hour 16:05 to 16:30 and again from hour 17:00 until delivery at hour 17:50, type II dips appear combined with tachycardia, indicating a severe fetal distress. Fetal blood sampling shows acidemia and hypoxemia which were confirmed at birth. Severely depressed newborn at the 1st, 5th, and 10th minute. The child is now 2½ years old and has a neurological deficit (the mother was breathing 100% oxygen most of the time).[30]

mother received an unusually high dose of Demerol (660 mg in twelve hours). This dose is large enough to produce pharmacologic depression of the newborn, even in the absence of true intrapartum fetal distress (acidemia, hypoxemia, hypercapnia). This interpretation is supported by the dramatic beneficial effect of the antidote to Demerol (N-allylmorphine), which was administered to the infant a few minutes after birth and caused rapid improvement of the condition of the newborn.

Recapitulation on the Prognostic Significance of Type II Dips

The preceding paragraphs demonstrate that type II dips should be considered as a sign of intrapartum fetal distress and as having an ominous prognostic connotation for the newborn.

A depressed newborn should be expected when during labor more than 20 type II dips have been produced, the sum of their amplitudes is greater than 600 amplitude units, and more than 35 percent of uterine contractions have produced type II dips.

The preceding statement has the following limitation: if type II dips are produced early in labor, and if they disappear at least two hours before delivery, the newborn may be vigorous. On the other hand, the presence of type II dips during the last two hours of labor is associated invariably with a depressed newborn, even if there were no type II dips in earlier stages of labor.

When type II dips appear, the obstetrician should investigate the causes of fetal distress and, if possible, correct them before irreversible damage is produced to the fetus. If this correction is not possible, the fetus should be delivered soon.

In the absence of type II dips the prognosis for the newborn is good. However, pharmacologic depression may be present in the absence of intrapartum distress if high doses of anesthetic or analgesics have been given to the mother during labor. Pure pharmacologic depression can be easily managed, and, if corrected, leaves no permanent damage to the infant.

Early Intrapartum Diagnosis of Fetal Distress

It is of utmost importance to recognize fetal distress during labor before irreversible damage has been produced to the fetal brain. Simultaneous graphic recording of FHR and uterine contractions enables the obstetrician to make such an early diagnosis, since the patterns corresponding to fetal distress are strikingly different from those obtained in normal conditions (Fig. 17). Several models of clinical monitors that produce graphic records

prenatal life

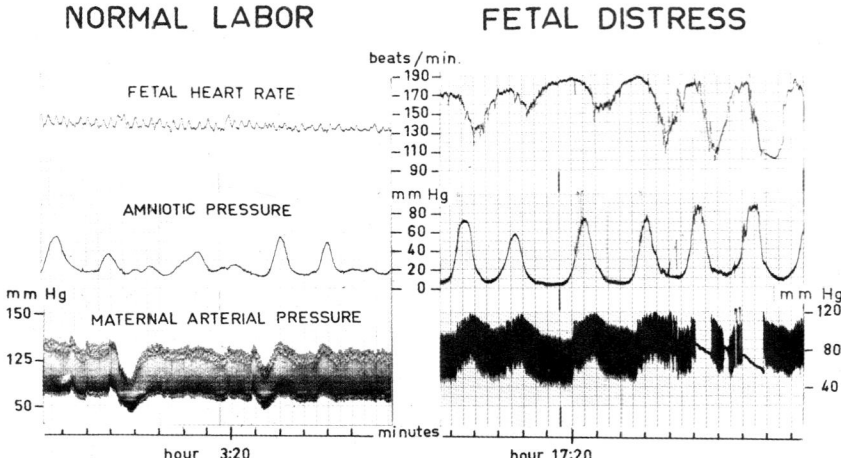

Figure VII-17. Typical patterns of FHR in labor. At the left is shown an example of the pattern usually recorded in normal labors delivering vigorous newborns. The Baseline FHR oscillates around 140 beats/min. Uterine contractions do not produce type II dips in FHR. At the right is shown an example of FHR pattern characteristic of severe fetal distress. The Baseline (Basal FHR) is abnormally high, ranging from 170 to 190 beats/min. Each uterine contraction causes a type II dip (transient fall in FHR occurring immediately after the contraction). At hour 17:10 sampling of fetal blood showed acidosis and hypoxemia. Forceps delivery at hour 17:53. The Apgar score was 1 at one minute and 2 at five minutes. This segment of a record belongs to case #1712; the corresponding perinatal chart is shown in Figure 16.[8]

Figure VII-18. Schematic illustration of the correct method for recognizing a type II dip by clinical auscultation of FHR. Fetal Heart Rate is counted during several consecutive periods of 15 seconds each. The average FHR corresponding to each period should be noted. For this purpose it is useful to rest for 5 seconds between every two counting periods.

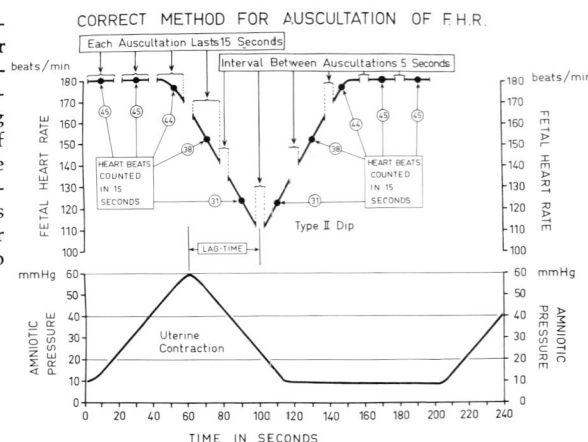

Correct Method for Clinical Auscultation of FHR

of the kind illustrated in Figure 17 are available at the present time. With such methods the diagnosis of the intrapartum condition of the fetus can easily be made.

Even in the absence of a graphic record of FHR and of uterine contractions it is possible to diagnose fetal distress by clinical auscultation of FHR if this is made following certain rules. The recognition of an FHR pattern, such as the one illustrated in the right side of Figure 17, can be made by simple clinical auscultation, provided that the fetal heart rate is counted during several consecutive periods of fifteen seconds each. At least twenty such periods should be counted (Fig. 18).

The count should start before the onset of a uterine contraction and should continue for at least two minutes after the uterus is fully relaxed. The average FHR corresponding to each period should be noted. For this purpose, it is useful to rest for five seconds between every two counting periods. The chronological relationship of each period with the uterine contractions should also be noted. The timing of the uterine contraction is made by manual palpation of the abdomen.

This method permits the reconstruction of an average curve of the variations of FHR and the relation of variations with the peak of the uterine contraction. Type II dips are perceived as a slow fall in FHR which starts after the peak of the contraction and reaches minimum values between thirty and fifty seconds later, that is, when the uterus is fully relaxed. The recovery of the FHR to its initial value is also slow and takes about the same time as the fall. In type II dips the changes in FHR are slow and, therefore, may not be perceived without actually counting and noting the rate every fifteen seconds.

The more or less stable values of FHR counted between the dips correspond to the Basal FHR; its value should be estimated as the average of all the counts made between the dips.

Any drop of fetal heart rate showing the chronological pattern described above should be considered as a type II dip. For example, if before the contraction FHR is 170 beats/min and thirty seconds after the peak of

prenatal life

the contraction it is 130 beats/min, the change should be considered as a type II dip with an amplitude of 40 amplitude units.

Incorrect method

If only one single auscultation of FHR is made between two contractions, it is very difficult to make the clinical diagnosis of fetal distress, even in a case having the typical FHR pattern illustrated in the right side of Figure 17. There are great probabilities that one single auscultation would find an average value between 160 and 120 beats/min, which would be considered as normal, although it may be a typical tracing of very severe fetal distress.

The wide clinical use of this incorrect method of auscultation may explain why several authors have not been able to find disturbances in FHR in fetuses undergoing severe distress, as revealed by the low pH of scalp capillary blood in utero and the marked depression of the newborn.

Summary

In order to establish the diagnostic and prognostic significance of certain patterns of Fetal Heart Rate (FHR) recorded during labor, we correlated these patterns with the condition of the newborn as evaluated by the Apgar score.

This study was made in forty-nine labors throughout which FHR was continuously graphically recorded by electronic methods. Uterine contractions were simultaneously recorded on the same paper.

Two elements of FHR tracings were found to have good correlation with the condition of the newborn: (1) The Baseline of FHR tracing (or Basal FHR), which in the vigorous group (Apgar score 7-10) had an average value of 143 beats/min, whereas in the depressed group (Apgar 1-6) it averaged 166 beats/min. The difference between both means is highly significant. A tentative limit has been set at 155 beats/min; and the rise of the Baseline above this limit is considered as a sign of fetal distress. The higher and the more prolonged the rise of the Baseline, the worse is the prognosis for the infant. (2) The type II dip, i.e., a transient fall of FHR caused by one uterine contraction and occurring after the contraction in such a way that the lag-time between the peak of the contraction and the bottom of the dip is thirty-fifty seconds. In the group with depressed newborns the average number of type II dips recorded during each labor was ninety-one whereas in the group with vigorous newborns the mean value was eleven type II dips. The difference between the two means is highly significant.

The authors consider that in normal conditions type II dips should be absent from the intrapartum record of FHR. The appearance of type II dips is a sign of fetal distress. However, if the total number of type II dips recorded is less than twenty, the fetal disturbance will not be severe enough to cause a low Apgar score.

When more than twenty type II dips have been recorded during labor, the newborn is usually depressed. The few exceptions to this rule were cases in which the type II dips were recorded in early stages of labor and then disappeared, the FHR tracing being normal for several hours before delivery.

The two abnormal signs in FHR (high Baseline and type II dips) usually appeared in association.

The continuous recording of FHR simultaneously with that of uterine contractions enables the obstetrician to make the early diagnosis of intrapartum fetal distress.

References

1. Apgar, V.: A proposal for a new method of evaluation of the newborn infant. *Anesth Analg* (Cleveland), *32:* 260-267, 1953.
2. Arnt, I. C., *et al.*: Seimiología do sofrimento fetal durante o trabalho de parto. Eighth Brazilian Congress Obstet Gynec, Recife, Brazil, Sept., 1966.
3. Berendes, H. W.: Intrapartum fetal heart rate and neurological development of children. In Perinatal Factors Affecting Human Development, special session in eighth meeting, Advisory Committee on Medical Research, Pan American Health Organization, Washington, June 1969.
4. Brady, J., and James, L. S.: Fetal heart rate and the onset of respiration. *Amer Pediat Soc Program and Abstracts.* Seventy-first Annual Meeting, Atlantic City, 1961.
5. Brady, J., James, L. S., and Baker, M. A.: Heart rate changes in the fetus and newborn infant during labor, delivery and the immediate neonatal period. *Amer J Obstet Gynec*, *84:* 1-12, 1962.
6. ———: Fetal electrocardiographic studies: Tachycardia as a sign of fetal distress. *Amer J Obstet Gynec*, *86:* 785-790, 1963.

7. Caldeyro-Barcia, Richard, et al.: Effects of abnormal uterine contractions on a human fetus. Mod Prob Pediat, 8: 267-295, 1963.
8. ———: Control of human fetal heart rate during labor. In *The Heart and Circulation in the Newborn and Infant*, ed. D. E. Cassels, New York, Grune & Stratton, 1966, pp. 7-36.
9. ———: Effects of abnormal uterine contractions on fetal heart rate during labor. *Supplementary Main Papers*, Fifth World Congress of Gynaecology and Obstetrics, Sydney, Butterworths, 1967, pp. 9-29.
10. ———: Intrapartum disturbances in fetal homeostasis and their correlation with respiratory distress syndrome and abnormal EEG in the child. In *The Child*, ed. A. Dorfman, Chicago, Yearbook Publishers, 1968, pp. 201-27.
11. Cox, L. W.: Foetal anoxia. *Lancet*, 1: 841-844, 1963.
12. Fenton, A. N., and Steer, C. M.: Fetal distress. *Amer J Obstet Gynec*, 83: 354-362, 1962.
13. Figueroa J. G., et al.: Las "espicas" de la frecuencia cardíaca fetal. *IV Congreso Uruguayo de Ginecotocología*, Montevideo, March, 1964, 2: 852-860.
14. Fitzgerald, T. B., and McFarlane, C. N.: Foetal distress and intrapartum foetal death. *Brit Med J*, 2: 358-361, 1955.
15. Ginsburg, S. J.: The significance of the signs of fetal distress: A preliminary study. *Amer J Obstet Gynec*, 74: 264, 1957.
16. Ginsburg, S. J., and Gerstley, L.: Fetal tachycardia in labor. *Amer J Obstet Gynec*, 92: 1132-1139, 1965.
17. Gulin, L. A., et al.: Sofrimento fetal: Fisiopatologia. Lecture presented at the *Seventh Brazilian Congress of Obstetrics and Gynecology*, Recife, Brazil, Sept., 1966.
18. Hon, E. H.: The electronic evaluation of the fetal heart rate: Preliminary report. *Amer J Obstet Gynec*, 75: 1215-1230, 1958.
19. ———: Observations on pathologic fetal bradycardia. *Amer J Obstet Gynec*, 77: 1084-1099, 1959.
20. ———: The fetal heart rate patterns producing death in utero. *Amer J Obstet Gynec*, 78: 47-56, 1959.
21. ———: The diagnosis of fetal distress. *Clin Obstet Gynec*, 3: 860-873, 1960.
22. ———: Detection of fetal distress. *Fifth World Congress of Gynaecology and Obstetrics*, Sydney, Butterworths, 1967, pp. 58-74.
23. Hon, E. H., and Quilligan, E. J.: The classification of fetal heart rate. II: A revised working classification. *Clin Obstet Gynec*, 11: 145, 1968.
24. Ibarra Polo, A. A.: Frequencia cardiaca en el sufrimiento del feto intraparto. (Unpublished thesis, University of Chile School of Medicine, 1967).
25. James, L. S., et al.: The acid-base status of human infants in relation to birth asphyxia and the onset of respiration. *J Pediat*, 52: 379-394, 1958.
26. Labo, G., Facci, M. and Valenti, G. V.: La sofferenza fetale: Rilievi clinici e grafici. *Riv Ital Ginec* 44: 381-406, 1961.

27. Lund, C. J.: The recognition and treatment of fetal heart arrhythmias, due to anoxia. *Amer J Obstet Gynec, 40:* 946, 1940.
28. ———: Fetal tachycardia during labor: A fallible sign of fetal distress. *Amer J Obstet Gynec, 45:* 626, 1943.
29. Méndez-Bauer, C., et al.: Effects of atropine on the heart rate of the human fetus during labor. *Amer J Obstet Gynec, 85:* 1033-1053, 1963.
30. ———: Relationship between blood pH and heart rate in the human fetus during labor. *Amer J Obstet Gynec, 97:* 530-545, 1967.
31. Moggia, A. V., et al.: Oscilaciones rítmicas de la frecuencia cardiaca fetal (FCF). *Fourth Uruguayan Congress of Obstetrics*, Montevideo, 1964, 2: 845-51.
32. Roszkowski, I., Kretowicz, Jr., and Wichrzycki, A.: Polygraphic fetal-heart-rate examinations in fetal distress. *Obstet Gynec* 22: 455-460, 1963.
33. Schifferli, P. Y.: La frecuencia cardiaca fetal a et largo del embarazo. Estudio de la influencia del sistema nervioso autónomo sobre el cronotropismo cardiaco fetal. *Archivos de Ginecología y Obstetrica,* 23: 87-97, 1968.
34. Steer, C. M.: The early diagnosis of fetal distress by clinical and electrocardiographic methods. *New York J Med,* 57: 1925, 1957.
35. Swartwout, J. R., Campbell, W. E., Jr., and Williams, L. G.: Observations on the fetal heart rate. *Amer J Obstet Gynec,* 82: 301-303, 1961.
36. Walker, J.: Fetal distress. *Amer J Obstet Gynec,* 77: 94-107, 1959.

chapter VIII

The "Ovariectomy Syndrome" and the Initiation of Labor

Arpad I. Csapo, M.D.

Introduction and Acknowledgment

Life is a broad, venturous word, reflecting harmony, beauty, and order. Accordingly, the biologist examines life with pure intellectual curiosity, sustained over the centuries by man's persistent desire to know.

When the adjective "prenatal" is added, fear, misery, and disease dominate the noun. The medical biologist, pressed by urgent obligations, struggles with them in a world of cold reality. Life is our gamble; prenatal life is the hazard of our children. However, pressure and urgency are poor promoters of progress. Advances in prenatal care depend upon bold and penetrating inquiries into the basic laws of reproduction.

In entering obstetric residency twenty-five years ago I hoped to witness the creation and promotion of life, controlled by such precision that a healthy newborn child can emerge in harmony, beauty, and order. Instead I became impressed by our ignorance in correcting nature's faulty designs, manifesting in malformations, abortions, prematurity, fetal distress, and death. The assumption that these errors in regulation are reflected in uterine function led me to the study of parturient patients (Fig. 1). However, the overwhelming complexity[1,2] characteristic of this high level of organiza-

Supported by USPHS Grants HD-01416, HD-01478, and 5-K6-HD-20,169, NIH, and The Sunnen Foundation.

prenatal life

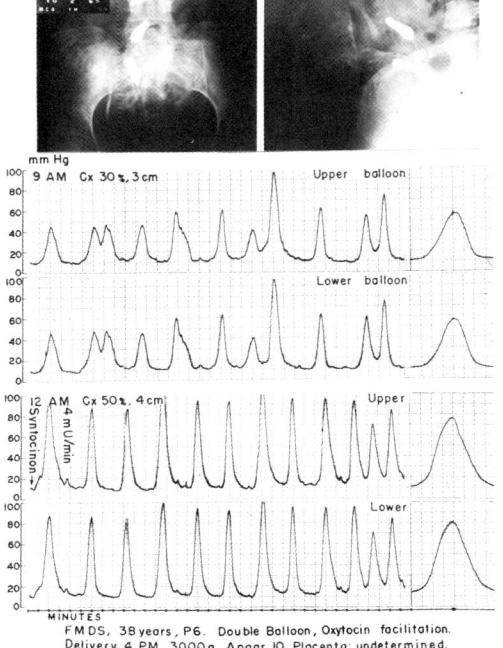

Figure VIII-1. The extraovular pressure of the human uterus during spontaneous and oxytocin facilitated labor. Double balloon recording; upper balloon above the presenting part, lower balloon between the ballotting head and the lower uterine segment (see X-ray evidence). Original record, illustrating that the intrauterine pressure at different points is similar. On the right, high speed recording. Note the increase in the rate of rise in pressure, on oxytocin administration. (Courtesy, A. Csapo: The placenta and the initiation of labor. *Nederl T Verlosk, 65:* 229-262, 1965.)

tion, obscured the regulatory principles I was looking for. Therefore, at this time it did not seem feasible to include in the studies an additional variable: the fetus. Basic laboratory research was needed for examining the regulation of uterine function step by step and at different levels of organization:[3] (1) the single myometrial cell in vitro; (2) the uterus as a whole in vitro, free of maternal control; and (3) the uterus together with this control in situ, in the intact animal (Fig. 2).

Persistence in systematic work, stimulated by conceptual and technical developments, permitted recently a study of fetal effects on the uterus.[4,5] For this limited interest the fetus paid off generously by displaying a powerful regulatory action on uterine function. By controlling the litter size surgically,[6] in the bicornuate rabbit uterus, we succeeded in controlling the initiation of labor predictably within the wide range of thirty to thirty-five days.

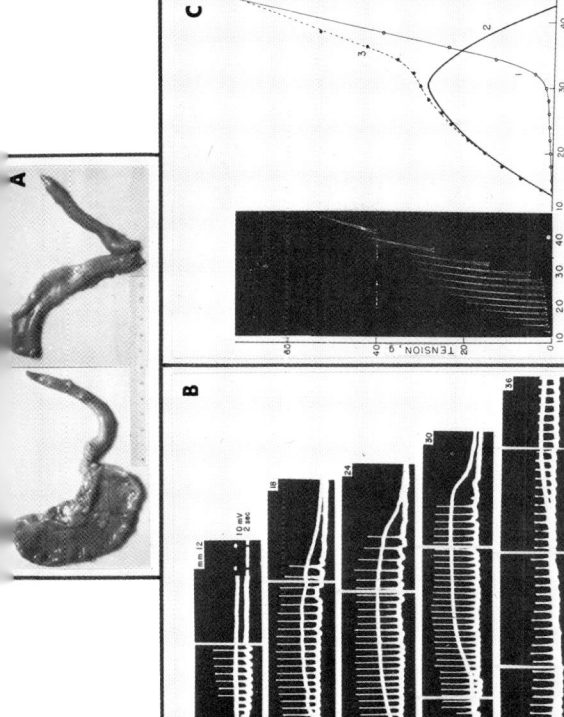
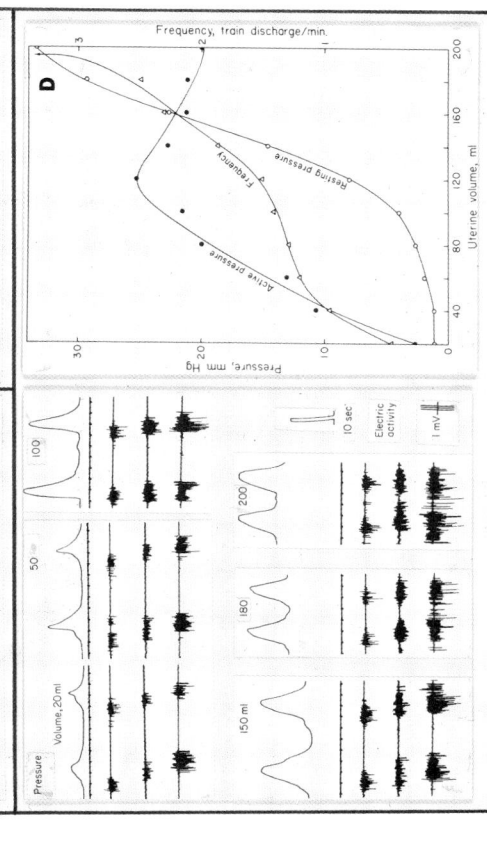

Figure VIII-2. The four cardinal effects of stretch on the myometrium. (a) Unilateral Hypertrophy, maintained (left) and induced (right), respectively, in postpartum and nonpregnant, ovariectomized rabbits by the chronic stretch (in the horn) of a recording balloon. (Courtesy, A. Csapo *et al.:* Stretch induced uterine growth, protein synthesis and function. *Nature* (London), *207:* 1378-13, 1965.) (b) The acute effect of stretch on the electric and mechanic activity of the excised uterus, in vitro. Original tracing; intracellular microelectrode technique. Note that a stepwise increase in uterine length increases the frequency and duration of train discharge and the isometric tension. Beyond the optimum length electric activity becomes continuous. (Courtesy, H. Kuriyama and A. Csapo, 1960.) (c) The length-tension relationship of the electrically tetanized excised rabbit uterus, in vitro. Original tracing (left) and graphical illustration (right) show the changes in "resting" (1) and "active" tensions (3), as well as the difference between the two (2), as a function of uterine length. Note that maximum active tension is recorded at the optimum length, i.e., before the sharp rise in resting tension. (Courtesy, A. Csapo: The mechanism of myometrial function and its disorders. In *Modern Trends in Obstetrics and Gynecology,* ed. K. Bowes, Butterworth, London, 1955.) (d) The acute effect of stretch on the electric and mechanic activity of the parturient rabbit uterus in situ. Original tracings (left), obtained by a extracellular macroelectrode assembly, and an intrauterine pressure receptor. Note the increase in the propagation velocity, frequency and the duration of train discharge and the increase in pressure with increasing volume; also the similarity in the length-tension (c) and volume-pressure (d) relationships. (Courtesy, A. I. Csapo, H. Takeda, and C. Wood: Volume and activity of the parturient rabbit uterus. *Amer J Obstet Gynec, 85:* 813-818, 1963.)

prenatal life

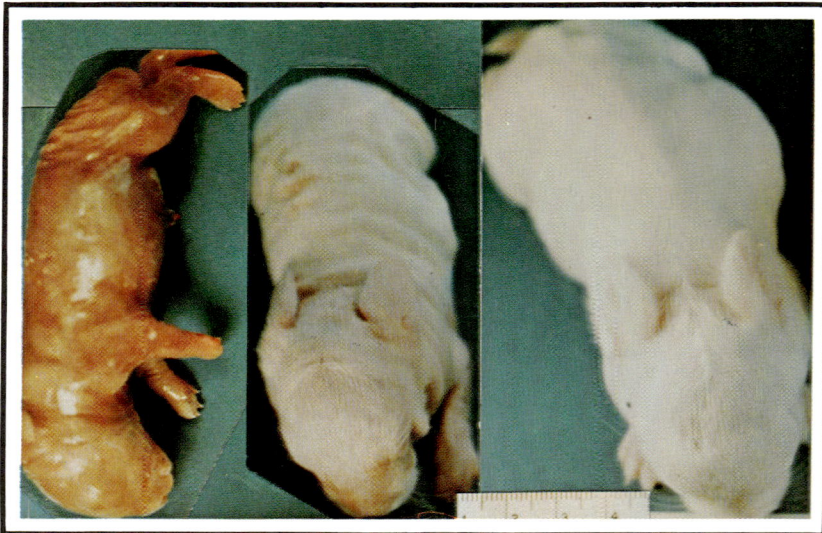

Figure VIII-3. The control of fetal birth weight and development in the rabbit, by the surgical control of litter size. Photographic illustration, rabbit newborns delivered by oxytocin at the thirtieth, thirty-third, and thirty-fifth days of pregnancy. The litter size, six, two, and one in one uterine horn was controlled surgically at day 25. Note the difference in size and stage of development of these three fetuses. (Courtesy, A. Csapo and E. Csapo, 1967.)

Fetuses of normal appearance and of normal 38 gram weight were delivered on the thirtieth day, and giants in strikingly advanced stages of development and of more than double size, 94 gram on the thirty-fourth to thirty-fifth day. In between these extremes were born the appropriate intermediates (Fig. 3).

It was at this stage, members of the committee, that your dynamic and provocative title, Prenatal Life, conspicuously attached to an invitation, reached us. I was challenged to return fulltime to the bench, the place for exploring without limitations in experimental design basic relationships between uterine function and the fetus. I wish to acknowledge your stimulus by presenting you with the "Ovariectomy Syndrome."

This syndrome is an experimentally induced endocrine deficiency, with closely predictable consequences on prenatal life. Under precisely controlled

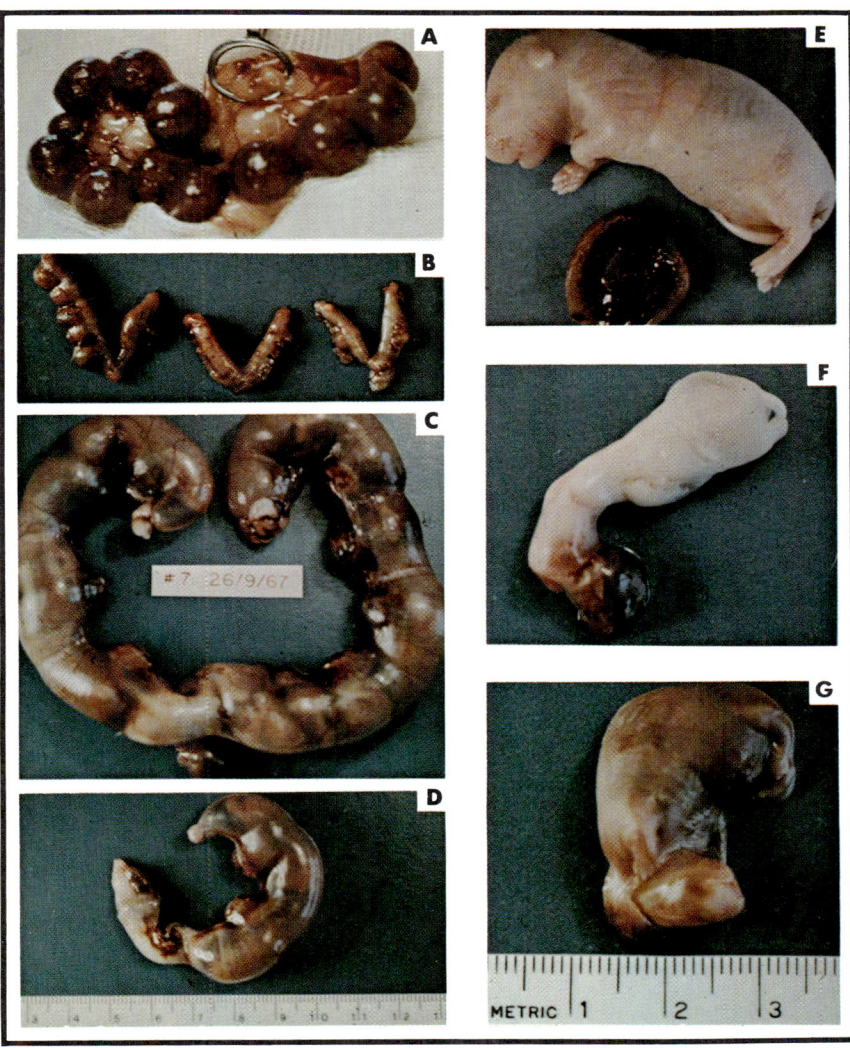

Figure VIII-4. The effect of ovariectomy in pregnant rats on fetal survival and development. (a) ovariectomy at day fifteen of pregnancy; (b) ovariectomy at days thirteen and fourteen, hysterectomy forty-eight hours later; (c) ovariectomy at day sixteen, hysterectomy at day twenty-one of pregnancy; (d) unilateral abortion after bilateral ovariectomy and unilateral placental dislocation at day seventeen; (e) fetus at day twenty-one delivered by hysterectomy from normal mother; (f and g) abnormal fetuses at day twenty-one, five days after ovariectomy. (Courtesy, A. Csapo and E. Csapo, 1967.)

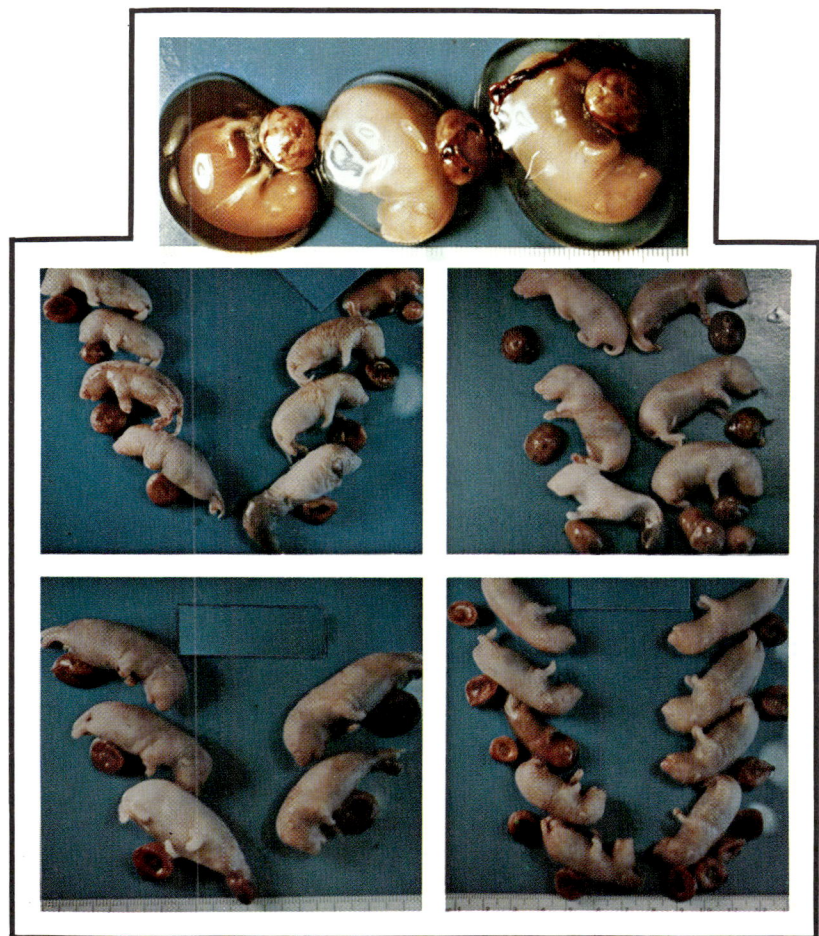

Figure VIII-5. Fetal defects, prematurity, and placental retention in ovariectomized rats. Hydramnios and four sets of litters are shown, delivered by hysterectomy at day twenty-one, five days after ovariectomy. Note the dead, deformed, discolored, as well as the macroscopically normal fetuses. Note also (right) placental retention, following partial premature delivery. (Courtesy, A. Csapo and E. Csapo, 1967.)

prenatal life

experimental conditions this deficiency, in the rat, results in a myometrial disfunction with a wide spectrum of prenatal accidents. These disorders are not only predictable, but are also readily preventable by a specific hormone therapy. The hormone harnessing uterine hostility and protecting prenatal life is, of course, progesterone, the steroid long since referred to as the "guardian of pregnancy."[7]

The Experiments

When prenatal defects manifest in gross deformities, hydramnios, discoloration, or death, the recognition and documentation of prenatal hardship is relatively simple (Figs. 4 and 5). What requires macroscopic scrutiny, combined with diagnostic histology, is the subtle anomaly of the mature living fetus. In order to promote the recognition of these fetal defects, the conceptus sites were examined, counted, and photographed not only at the time of ovariectomy but also at hysterectomy near term. The classification of these fetal anomalies, induced by ovariectomy, ranging between wide extremes, such as rapid death within twenty-four hours, followed by autolysis, absorption, and loss of uterine volume, on the one hand, and term pregnancy with macroscopic, microscopic, or no defects of the mature, viable fetus, on the other, will require several months of systematic histological work.

However, the microscopic examination of a number of normal and defective term fetuses by Richard Kempson, advanced the opinion that:

In areas with gross abnormalities there is marked degeneration typical of autolysis. There is no sign of inflammatory infiltration. All organs are formed, and are clearly recognizable, except in cases of early death and in those instances when whole extremities, or parts of the body are completely missing. However, in abnormal areas there is edema, disruption of organ architecture, and necrosis. This degenerative process is recognized at greater or lesser extent in the different tissues of the different fetuses, i.e., in the nervous system, muscle, skin, and gastrointestinal tract. The hemorrhagic areas contain large numbers of red blood cells, in addition to degenerative tissues.

The Normal Controls

Our physiological studies began with the examination of normal, Holtzman bred, Sprague-Dawley pregnant rats. Between the twelfth and seventeenth

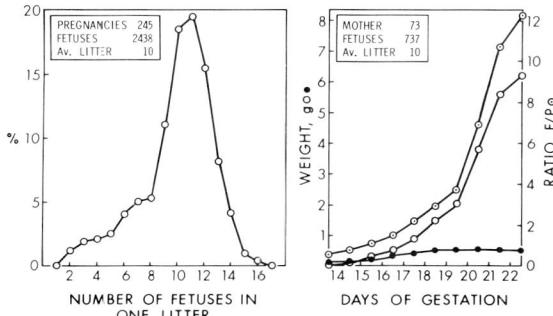

Figure VIII-6. The distribution of litter size; fetal and placental weights during pregnancy in the rat. Note that over 90 percent of the animals carry six to fourteen fetuses. Note also that placental growth decelerates after the nineteenth day of gestation, while fetal growth accelerates. Note the marked increase in the F/P ratio after the nineteenth day. (Courtesy, A. Csapo and E. Csapo, 1967.)

day of gestation 245 animals were operated under ether anesthesia and strictly sterile precautions. A total of 2,438 conceptus sites were counted and graphically recorded during laparotomy. The great majority, 80 percent of the animals, had an asymmetric distribution of the litter in the two uterine horns. The total litter size varied in such a way that 94 percent of the animals had six to fourteen fetuses. Apparently a litter of sixteen is a biological upper limit, for no animal carried more than sixteen fetuses (Fig. 6).

In seventy-three normal controls, 737 fetuses and placentae were removed by hysterectomy, between the twelfth to twenty-second day of gestation. Using these specimens the normal fetal and placental growth and the fetal/placental weight ratios were determined. The similarity in fetal and placental weights was striking between members of the same and different litters of similar gestational age, yielding stable fetal/placental ratios (Fig. 6).

The Holtzman Farm with a breeding experience of 30,000,000 animals offered the following explanation for this stability in fetal and placental growth. The animals were maintained at constant temperature and diet. They were selected for our experiments when eighty to ninety days old, 200-230 g weight, dated with plus or minus three hours accuracy on day zero, by sperm positivity. We were advised that these animals deliver spontaneously in about thirty minutes on the twenty-second day, as a rule, exceptions being mothers with unusually small or large litters. We were also informed that the fetal anomalies illustrated by Figures 4 and 5 were not seen among the newborns of normal animals. However, we noted that defective fetuses born at the laboratory from an ovariectomized mother may be eaten by her

prenatal life

and thus remain unnoticed. This is why we examined the near-term litter during hysterectomy and allowed only eleven animals to deliver spontaneously. In this group we observed no defective fetuses. Out of 245 normal control animals, observed during laparotomy between the twelfth to twenty-second days of pregnancy, only seventeen mothers carried one or more visibly defective fetuses.

The Effect of the Time of Ovariectomy on the Maintenance of Pregnancy and Fetal Survival

In contrast, the animals that were ovariectomized under strictly sterile precautions between the twelfth and seventeenth days of gestation invariably carried a varying number of dead or defective fetuses. Depending on the gestational date of ovariectomy, the litters responded differently to the suspension of luteal function. This is evidence that prenatal life depends strongly on the timing of ovariectomy.

Ovariectomy at the twelfth or thirteenth day of gestation resulted in death within twenty-four to forty-eight hours, followed by autolysis, absorption, and the reduction in uterine volume. Ovariectomy at the fourteenth day resulted in either the same, or, as did ovariectomy between the fifteenth to seventeenth days, it led to the continuation of pregnancy and premature delivery with dead, deformed, partly destroyed, or living fetuses. Term pregnancy was reached only by animals ovariectomized on the fifteenth day or thereafter, part of the litter having been delivered prematurely.

The term fetuses had macroscopically distinct defects of greater or lesser degree or appeared normal (Figs. 4 and 5). These apparently mature and viable term fetuses are unique in that their prenatal life was mortally endangered by the withdrawal of luteal support. It is of interest to note that in the human the luteoplacental shift in progesterone support normally occurs at the end of the first trimester and is complete.

A conservative estimate based on the direct observation of the great majority of the litters shows (Fig. 7) a sharply increasing rate of fetal survival if ovariectomy is done after the fourteenth day. Ovariectomy at the fifteenth to seventeenth days yields litters (or partial litters) of apparent vitality and maturity. This is evidence that the animal engages in considerable efforts aiming at the compensation of the experimentally induced hormone deficiency. These efforts are only partly successful, as shown by the variable success of fetal survival, variations in fetal development, and in viability.

Fetal survival in ovariectomized animals is directly influenced by allowing the mother to reach the twenty-first day without intercepting pregnancy by hysterectomy. Exceptions are instances in which early fetal death in utero is diagnosed by palpation. Such experiments showed that ovariectomy before the thirteenth day (in ten animals) resulted in invariable death of the entire litter. Of ten animals ovariectomized on the fourteenth day only two reached the twenty-first day, both in advanced premature labor. Of nine animals ovariectomized on the fifteenth day three reached the twenty-first day in early premature labor. Of twenty-one animals ovariectomized on the sixteenth and seventeenth day twenty reached the twenty-first day, two in advanced, six in early premature labor, and twelve with complete litters.

Theoretical Considerations

This strong dependence of the maintenance of pregnancy, fetal development, and survival on the time when luteal support is withdrawn is of considerable interest (Fig. 7). On day thirteen the average placental weight of the normal

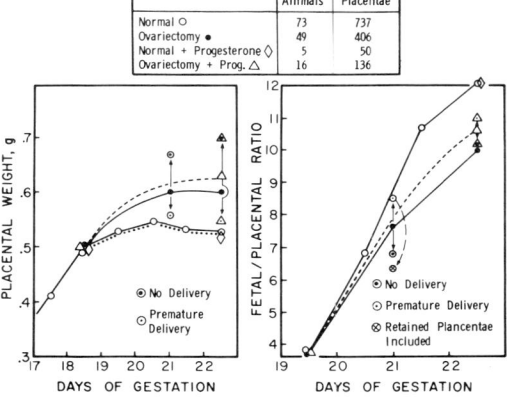

Figure VIII-7. The effect of the time of ovariectomy on pregnancy and fetal survival. Since not all fetuses were recovered, premature delivery was considered of constant duration at twenty-four hours. Note that the later ovariectomy is performed, between the thirteenth to seventeenth day, the longer pregnancy is maintained and the greater is fetal survival (solid lines). Broken line: fetal survival, at twenty-first day, as a function of the time of ovariectomy. Note the position of this curve (broken line) on the gestational time table and compare it with placental growth (Fig. 6). (Courtesy, A. Csapo and E. Csapo, 1967.)

prenatal life

animal is only 0.11 g, a value that quadruples during the subsequent four days. During this same period the success of ovariectomized animals in sustaining pregnancy sharply increases. The possible regulatory significance of this increase in placental weight demanded theoretical considerations and detailed investigations. These results were not predicted by a considerable, but unusually controversial, literature,[8-11] difficult to review briefly. What seemingly has not been fully recognized by most investigators was the significance of the: (1) time of ovariectomy with respect to gestational age; (2) improved surgical techniques, promoting the predictability of the results; (3) the counting, weighing, and photographing of the fetuses and placentae at ovariectomy and at spontaneous delivery, or at hysterectomy; (4) macroscopic scrutiny, combined with diagnostic histology, in evaluating the fetus; and (5) the inability of rats, ovariectomized between the fifteenth and seventeenth days of pregnancy, to sustain their litter until term during the summer season and when in imperfect health (e.g. slight diarrhea).

In our earlier pilot experiments,[12] while focusing upon the electrophysiological effects of ovariectomy on the uterus, we nearly repeated the same errors. Although we noted, as did others,[8-11] that ovariectomy after the fifteenth day of gestation does not invariably terminate pregnancy, that premature delivery is frequently associated with placental retention, that massive progesterone therapy maintains pregnancy, and that parturition is abnormal and the lowermost fetus is often deformed, we temporarily overlooked the underlying rules, reflecting regulatory principles.

Subsequent experiments were greatly aided by recent advances in our conceptual framework of the mechanism of myometrial function and regulation. Reference is made here to a series of comparative studies, examining the "evolution" of uterine activity in 132 pregnant rabbits[13] and ninety-six patients,[14] complementing earlier studies of others[15,16] and our own. While the regulatory significance of uterine volume and progesterone remained insufficiently documented by these studies we acquired new ways of looking at an old, complex problem. These experiments developed the concepts of the "systemic" and "local" progesterone block and the resulting endocrine and functional "symmetry" and "asymmetry" of the uterus; the block; the regulatory significance of the uterine volume/progesterone ratio (V/P); the distinction between progesterone compensatory "reserve" and "effort" of the placenta, etc. Since the experimental background leading to these concepts are described in two recent comprehensive reviews,[4,5] they need not be discussed here.

In pregnant rabbits the luteal component is the major factor controlling pregnancy, and a placental component under physiological conditions is only barely indicated. In second and third trimester patients the luteal component is no longer detectable, and a penetrating study of the placental component in rigorously controlled experiments is seldom permissible and feasible.

Thus, a readily accessible species was needed for analytical inquiries, an animal with such a luteoplacental compromise in which the contribution to the progesterone block of the two components varied with gestational age, or could be altered experimentally. The experiments so far described suggested that pregnant rats might be the desired experimental animals, suitable for the documentation of the regulatory significance of uterine volume and progesterone, and also for the local and systemic components of the progesterone-control mechanism.

Placental Hypertrophy Induced by Ovariectomy

We assumed that pregnancy in rats is maintained by a luteoplacental compromise and that the contribution of the two components (luteal and placental) to the progesterone block, as well as the placental compensatory potential, changes with gestational age. If so, then the maintenance of pregnancy and fetal survival must depend sharply upon the time of gestation, when luteal support is suspended. We demonstrated that this is indeed the case (Fig. 7).

If after ovariectomy the maintenance of pregnancy and fetal survival depend upon the effective support of the progesterone block by the placenta, and if the compensatory reserve of the placenta is minimal, then the placenta must undergo endocrine hypertrophy during a compensatory effort. We considered the possibility that this endocrine hypertrophy has a structural manifestation.

Therefore, placental weight was measured at different times after ovariectomy (Fig. 8, Left). The experiments showed a placental weight increase in comparison with the controls. A significant additional point seemed apparent when the placental weights were considered in the ovariec-

prenatal life

tomized animals that sustained pregnancy beyond the nineteenth day (Fig. 8). The animals could be divided into two groups: (1) in premature labor; and (2) with no signs of labor. It then became apparent that success in the maintenance of pregnancy depends upon placental hypertrophy. Animals in premature labor showed no placental hypertrophy (on the average), while those not in delivery did. These observations suggest that placental hypertrophy is aimed toward an endocrine hypertrophy.

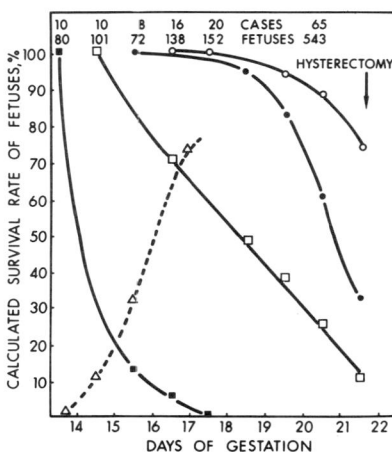

Figure VIII-8. The effect of ovariectomy and decremental progesterone treatment on placental growth and on the F/P ratio. Left: The average increase in placental weight, following ovariectomy (black dots) and ovariectomy plus decremental progesterone treatment (interrupted line) is compared with that of the controls (dotted line). The difference in placental weight at the twenty-first day between prematurely delivering and non-delivering animals is also shown. Right: Note, that the F/P ratio is lowered by ovariectomy (black dots), due to placental hypertrophy. Note that decremental progesterone treatment following ovariectomy, lowers the F/P ratio in spite of increased fetal weight and that the F/P ratio is the lowest when fetuses are delivered but their placentae are retained. (Courtesy, A. Csapo and E. Csapo, 1967.)

Premature Delivery and Placental Retention

Ovariectomized rats utilized another rather ingenious compensatory mechanism in the promotion of pregnancy. This mechanism may be looked upon as a last resort in effective compensation. Instead of discharging the premature fetuses in rapid succession (as normal controls deliver the term litter), ovariectomized rats expelled fetuses several hours or even several days apart, without delivering their placentae (Fig. 5). With the delivery of each fetus, not followed by the placenta, the fetal/placental ratio was lowered markedly (Fig. 8). The delivery of each fetus lowered the uterine volume. This unloading of the uterus, combined with placental retention, permitted the parturient animal to interrupt the delivery process and remain pregnant for long subsequent periods. Delivery was resumed when the uterine volume increased by uninterrupted fetal growth.

The fetal/placental ratio was found to be very stable in normal controls (Fig. 8). Ovariectomized animals showed a lower (average) ratio than the controls. However, the ovariectomized animals which were in premature delivery simulated the controls, while those which were not had a lower F/P value. The fetal/placental ratio of animals in premature delivery was the lowest when the retained excess placentae were also computed. This corrected lowermost value is a strong indication that the maintenance of pregnancy in ovariectomized animals is dependent upon the fetal/placental ratio. Decremental progesterone treatment did not affect the F/P ratio in the normal animals, but lowered significantly the ratio in the ovariectomized animals. This was the case in spite of increased fetal weight, after progesterone therapy.

The Luteoplacental Shift

A luteoplacental shift in progesterone support, induced in pregnant rats by ovariectomy, is strongly indicated by the observed placental hypertrophy and by the promotion of hypertrophy after decremental progesterone treatment. Further support is provided by the observed relationship in the ovariectomized animal between placental hypertrophy and the success of the maintenance of pregnancy. However, steroid chemical evidence would strengthen documentation considerably and promote the concept that this shift involves a change from a systemic to a local progesterone effect. The physiological evidence predicts that after ovariectomy pregnancy is maintained, because through placental hypertrophy the total placental progesterone content increases, a change which sustains to some extent the myometrial progesterone concentration. This otherwise would be drastically reduced due to a drastic reduction in peripheral plasma progesterone, which can be partly restored by progesterone substitution therapy.

In attempting this documentation blood, placenta, and uterine tissues were collected under precisely defined physiological conditions. The steroid chemical study of these samples was conducted by Dr. Walter Wiest, using his sensitive and specific "double isotope" technique.[17] This study showed

prenatal life

(Fig. 9) that peripheral plasma progesterone drops not only far below the normal but even below the parturient value in ovariectomized rats; that this level is increased by progesterone treatment; that the total placental progesterone content increases; and that the progesterone concentration in the uterus does not change proportionately with the blood levels.

Since these tentative conclusions are based on pilot experiments, more extensive studies are desired and are in progress. However, in view of the additional electrophysiological evidence, to be presented, complementing the physiological and chemical data, a tentative conclusion would seem permissible. It appears that a critically timed ovariectomy in a pregnant rat does induce a luteoplacental (and systemic—local) shift in progesterone support. This shift sustains the myometrial block at an extent demanded by the continuation of pregnancy in spite of drastically reduced blood progesterone levels. The varying success in the promotion of pregnancy and fetal

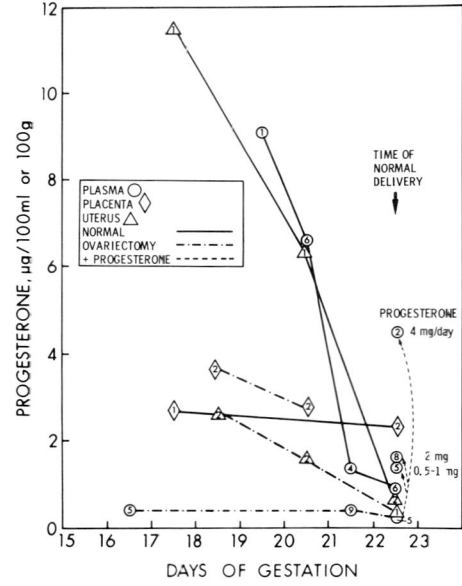

Figure VIII-9. Progesterone concentrations in plasma, placental, and uterine tissues of normal and ovariectomized pregnant rats. "Double isotope" method. Numbers in symbols represent the number of animals providing the samples. Blood was collected by heart puncture and tissue by hysterectomy. Bilateral ovariectomy was carried out at fifteen to sixteen days of pregnancy. Progesterone was given in oil i.m., twice a day. Note the withdrawal in plasma and uterine progesterone in the normal control animals; the low plasma and relatively high placental and uterine concentrations after ovariectomy; and the increase in plasma progesterone of ovariectomized animals after hormone substitution therapy. (Courtesy, W. Wiest and A. Csapo, 1967.)

survival indicates the delicacy of this profound regulatory change. However, undisturbed pregnancy and fetal development are fully guaranteed by progesterone replacement therapy and even by effective partial replacement in the form of decremental progesterone treatment. As will be documented presently, the success of decremental therapy in guaranteeing normal pregnancy depends upon the success of this treatment in inducing marked placental hypertrophy. To the best of our knowledge, these experiments represent the first attempt in the resolution of a basic regulatory condition through the combined techniques of physiology, electrophysiology, steroid chemistry, and histology.

The V/P Ratio

The striking effect on uterine activity of an increase in uterine volume, promoting abortion or labor, has been documented in a great variety of physiological, electrophysiological, and clinical studies.[4, 5] It is of interest here to recall that in cases of fetal death in utero, when a mere increase in uterine volume effectively triggers uterine activity and abortion,[5] the plasma progesterone was found to be far below normal.[17] Volume Induction was ineffective, however, during normal midtrimester pregnancy.[5] An interesting exception from this rule was offered by a clinically normal thirteen weeks pregnant patient who was successfully induced by volume increase. However, this patient had abnormally low plasma progesterone values, a finding which strengthens the rule (Fig. 10).

The experiment that follows is the most simple form of additional documentation for the volume control of the initiation of labor. Direct, visual observation and graphic and photographic recording during ovariectomy revealed an asymmetric distribution of conceptus sites in the two uterine horns in 80 percent of the animals. Near term the fetal and placental positions in the two uterine horns were examined and photographed during hysterectomy in twenty-six rats, which were in premature labor. It then became apparent that only two animals delivered first from the lesser distended horn, documented by the intrauterine position of the retained placentae. In all the others one or more retained placentae (without fetuses) were found attached in the more distended horn. However, in the two cases of exception, the conceptus sites in the two horns were nearly symmetrical, three and four, and four and five fetuses, respectively.

These observations show that in ovariectomized rats (lacking luteal progesterone support) the onset of premature labor is not systemically but locally controlled. Since the fetal/placental ratios are similar in the two

prenatal life

uterine horns when labor begins but the uterine volumes are not, the excess volume is suspected for the "weakening" of the block and for the initiation of premature labor. A systemic progesterone "withdrawal," triggering premature labor, cannot occur, for it already had when the ovaries were removed four to six days earlier. After the discharge of each fetus and the retention of each placenta the animal lowered the V/P ratio and interrupted premature delivery, as a rule, for hours or days. Premature labor was re-initiated when uterine volume increased again, due to uninterrupted fetal growth.

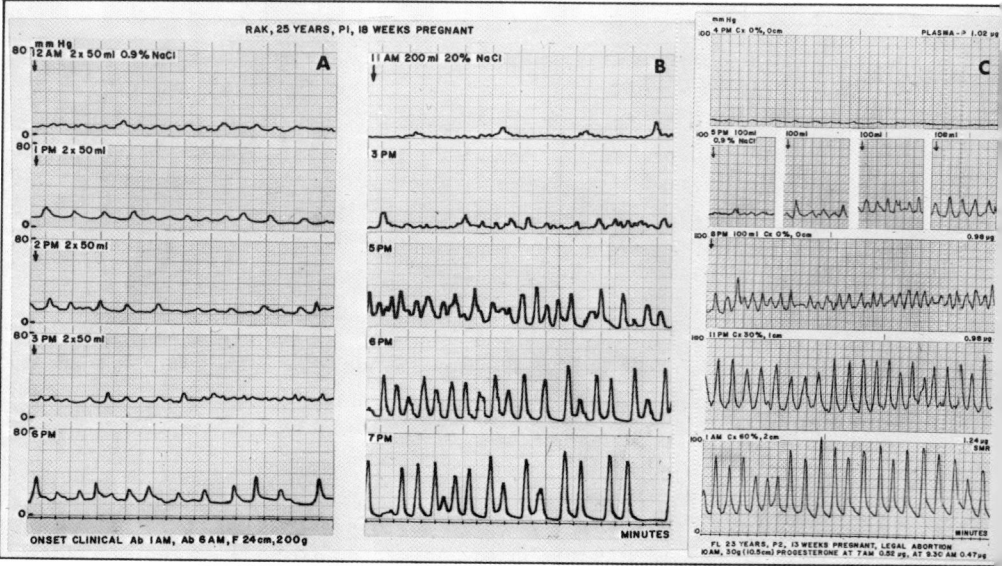

Figure VIII-10. The effect of uterine volume increase during normal midtrimester human pregnancy. (a) Moderate and transient increase in uterine activity, induced by the intraamniotic injection of 400 ml isotonic saline. This effect is contrasted (b) with the predictable accelerated evolution of uterine activity, induced (twenty-four hours later in the same patient) with hypertonic saline. Note the timing of induced abortion. (c) An exception from the rule, illustrated at (a). In this instance isotonic saline did induce the evolution of uterne activity and abortion. However, this clinically normal patient, like those carrying a dead fetus, had abnormally low progesterone values (see right corner of each tracing). (Courtesy, A. Csapo, M. Pulkkinen, J. Sauvage, and W. Wiest, 1967.)

The Initiation of Normal Delivery in the Rat

This proposed, primarily volume-controlled mechanism for the initiation of premature labor in ovariectomized rats does not explain the precisely timed and rapid delivery of the normal term animal. In spite of the fact that 94 percent of the normal controls had a sizable volume (carrying six to fourteen fetuses), more variation could be expected in both the timing and the duration of labor if labor were controlled primarily by volume. This assumption is based on the pronounced volume asymmetry in the two uterine horns of normal pregnant rats. Since the normal term animals deliver rapidly and with little variation in the time of the onset, we assumed that normal parturition in rats is dominantly controlled by a true, systemic progesterone withdrawal, only aided by a volume effect. It is of interest, therefore, that Walter Wiest and his associates[18] documented in rats by using the double isotope method a sharp decrease of peripheral plasma progesterone on the twenty-first day.

Thus in the same species the same basic mechanism controlling the initiation of labor may be slightly modified. An experimental modification is the reported alteration of the luteoplacental progesterone support, induced in the rat by ovariectomy. These animals can serve as models for the study of the initiation of labor in pregnant patients, who undergo a similar shift during first trimester pregnancy. This early shift permits ovariectomy in patients at the end of the first trimester without consequences on the maintenance of pregnancy and prenatal life. However, because of this early shift and the presence of only one placenta, placental defects may create regulatory disturbances, seriously threatening prenatal life.

Progesterone Substitution Therapy

If the above conclusions are essentially correct, then the restitution of the luteal component in ovariectomized rats, in the form of systemic progesterone substitution therapy, should not only guarantee the maintenance of pregnancy, but should also secure undisturbed prenatal life. Thus, a group of ten rats ovariectomized on the fifteenth or sixteenth days of gestation were treated daily with 2 to 4 mg progesterone, i.m., in oil, until day twenty-one. When hysterectomy was performed on the twenty-second day, normal and intact pregnancy was observed in all instances. An additional group of four

prenatal life

animals received a daily dose of 2 mg progesterone until day twenty, and these animals also had normal pregnancy at day twenty-two. This much is certain.

Three additional animals receiving only 1 mg progesterone per day also appeared to be normal. However, these fetuses must be subjected to diagnostic histology and additional experiments, because eleven animals that received a daily dose of only 0.5 mg progesterone, or up to 2 mg progesterone, but only two to four days after ovariectomy had either premature delivery or defective fetuses or both. Apparently the systemic progesterone requirement of ovariectomized rats is high, about the same as that of ovariectomized rabbits, animals twenty times larger. Also, the timing of treatment is important, as indicated by the failure of other observers to maintain pregnancy at this reported dose level. These observations are of considerable theoretical and practical significance, because of the persistent difficulties in documenting the myometrial blocking effect of exogenous progesterone therapy, when the pregnant patient is under the endogenous, intrauterine, control of the placenta. In contrast, this effect is readily demonstrated when nonpregnant women are under the extrauterine control of the ovaries.[5]

Compensatory Placental Hypertrophy Promoted by Decremental Progesterone Treatment

Having observed the relatively high progesterone demand for the maintenance of pregnancy of ovariectomized rats, we have assumed that the dose of progesterone might be both lowered and tapered off, if a compensatory placental hypertrophy is promoted by decremental progesterone therapy. This decremental treatment is aimed at mimicking the gradual progesterone withdrawal of the regressing corpus luteum of human pregnancy.

In setting the timetable of this experiment, we took implantation rather than conception as time zero, since only then does the placenta begin to develop. A period of five days is only 1/55th of the 280 days of the human gestation period, while it is less than one-fifth of the twenty-two-day-long rat gestation period. The significance of timing in this experiment is well reflected by the weight of the rat placenta eight days after implantation, a mere 0.11 g on the thirteenth day of pregnancy. The question we posed was: does decremental progesterone therapy, initiated as early in pregnancy as the twelfth to fourteenth day (seven to nine days after implantation) and continued until the seventeenth to nineteenth day, provoke placental compensatory hypertrophy? If so, to what extent does this compensatory effort promote pregnancy and prenatal life?

Table 1. The Effect of Decremental Progesterone Treatment in Control and Ovariectomized Pregnant Rats

Group Progesterone Days Treatment	No. Animals Fetuses	Total Average Fetal/Placental Weight, g	More than 1 Abnormal Intact, g	Premature, Percentage Dead, Percentage
Control 10.5 mg			—	—
15-18	5/50	6.7/.54	6.7/.54	—
14 mg			2.5/.47	38
12-18	12/128	3.1/.57	6.2/.98	79
12 mg			4.6/.53	27
13-18	26/276	5.5/.58	7.0/.72	51
6.5 mg			5.3/.50	27
14-17	4/37	5.5/.58	6.4/.85	43
10 mg			5.8/.58	0
14-17	5/49	6.2/.59	6.4/.62	10
12 mg			5.7/.58	0
14-19	16/162	6.7/.68	6.9/.70	5

The results obtained in sixty-eight ovariectomized rats, carrying 702 fetuses, are tabulated in Table 1. It is significant that the normal controls (not ovariectomized) receiving a total dose of 10.5 mg decremental progesterone showed no placental hypertrophy, in contrast to the ovariectomized animals which did. Evidently placental hypertrophy is provoked by ovariectomy.

The table illustrates that all groups of ovariectomized animals responded by placental hypertrophy (at various degrees) to the combined effect of ovariectomy plus decremental progesterone treatment. The delicacy of this experimentally provoked compensatory process is illustrated by the sharp dependence of its success on the timing of the procedure and the total dose of progesterone used. The difference between twelve to fourteen days

prenatal life

ovariectomy manifested in differences in the frequency of prematurity (38 and 0 percent) and fetal death in utero (79 and 5 percent, respectively). The strong dependence of success in fetal survival on the dose and duration of progesterone treatment is illustrated by the group of rats ovariectomized on day fourteen of pregnancy. An increase in the total dose of progesterone from 6.5 to 12 mg, prolonging treatment by one or two days, decreased the frequency of prematurity from 27 to 0 percent and fetal death in utero from 42 to 5 percent.

Evidently, properly timed ovariectomy plus decremental progesterone therapy at the effective dose level readily provokes placental compensatory hypertrophy. In so doing it promotes pregnancy and prenatal life. The great significance of ovariectomy is illustrated by the finding that in spite of progesterone treatment in the intact control animal (with functional corpora lutea), there is no hypertrophy. On the other hand, if ovariectomy is done too early, or the decremental dose is too low, hypertrophy is minimal. The importance of placental hypertrophy is further stressed by the varying response of individual animals to one and the same procedure. Within the same group the animals that responded to the procedure with significant compensatory hypertrophy maintained normal pregnancy with normal litter, whereas those with minimal or no hypertrophy did not. A good example of this individual variation is the single animal in a group of twelve, which like the other eleven animals received a total dose of 14 mg progesterone following ovariectomy at day twelve. In this animal, unlike the others, the average weight of eleven placentae was 0.98 g, in contrast to the 0.56 g total average of the group (128 placentae). This animal was the only one in the group that carried all, i.e. eleven normal fetuses to term, averaging 6.2 g. Thus, the high prematurity, or fetal mortality rate, in certain groups is no evidence that these animals were unable to perform the luteoplacental shift in the promotion of pregnancy. These failures only illustrate the delicacy of the procedure and that of the compensatory response.

Notably, several attempts at inducing delivery, with a single i.m. dose of 0.2-0.5 I.U. oxytocin, failed in those ovariectomized animals which as a result of placental hypertrophy carried an intact litter to term. This indicates that ovariectomy and the luteoplacental shift profoundly changes the regulatory conditions of the uterus.

These experiments illustrate that the luteoplacental shift is not a monopoly of the human species. The rat is capable of performing it if a compensatory placental hypertrophy is provoked nine days after implantation; ovariectomy and pregnancy is transiently supported by a gradual progesterone withdrawal. Evidently the sharp division of species, on the basis of their different response to the drastic intervention of abrupt and warningless ovariectomy, was ill-conceived. This test only demonstrated the relative progesterone contribution of the placenta in different species in the presence of a flourishing corpus luteum effect. At the same time this arbitrary division obscured the fundamental law that both man and laboratory mammals possess a placental compensatory potential, which is readily realized, by ovariectomy and a transient gradual withdrawal of systemic progesterone, a physiological process during first trimester human pregnancy, triggering placental compensatory effort.

These considerations led us to inquire whether the rabbit, the notorious species in which, in contrast to man, the ovaries were said to be "indispensable" for the maintenance of pregnancy, is also capable of performing the luteoplacental shift and compensatory placental hypertrophy. The significance of this study lies in the persistent claim of a basic species difference between the rabbit and man in the mechanism of the maintenance of pregnancy. The experiments in the rabbit showed that pregnancy can be sustained after ovariectomy, carried out as early in gestation as the eighteenth day (eleven days after implantation). These animals when receiving decremental progesterone treatment during ten days, totalling only 17-21 mg and tapering down to 0.5 mg during days twenty-six to twenty-eight of pregnancy, readily reached term, as a rule. Such animals delivered at or near term fetuses, averaging 68 plus or minus 1.2 g and placentae averaging 9.1 plus or minus 0.3 g, both about 60 percent higher than the normal values. In rabbits, as in rats, the success in the promotion of pregnancy after ovariectomy plus decremental progesterone therapy depended upon compensatory placental hypertrophy. Animals of normal placental weight delivered at the cessation of treatment, or shortly after. This is evidence that the success of the experiment depends upon the success of induced placental compensation and not on the "depot" effect of progesterone.

Evidently even the rabbit readily performs the regulatory phenomenon of compensatory placental hypertrophy, if effectively provoked by ovariectomy and supported by decremental progesterone therapy. This regulatory change operates in man, normally provoked by the regression of ovarian

prenatal life

function during first trimester pregnancy. Therefore, these experiments, complementing those performed in rats, bring into focus a basic regulatory law: the compensatory placental hypertrophy, provoked by ovariectomy, aims at the promotion of pregnancy and prenatal life.

An additional finding of interest is that while the average fetal weight of control rats was 6.2 g (at day twenty-two), the average weight of 123 fetuses supported after ovariectomy by sustained progesterone therapy was 7 g, and that of the 214 fetuses aided by decremental therapy was 6.7 g. Apparently progesterone therapy increases fetal weight.

The Local Effect of the Placenta on Uterine Activity

The concepts, of the "local" effect of placental progesterone, the endocrine and functional "asymmetry" of the uterus, and the regulatory significance of the V/P ratio, resulted from the recognition of the significance of nature's own experiments.[1,5] Reference is made here to those instances when human twins are born several months apart from the same uterus. Since the rat (like the human) is able to sustain pregnancy without luteal support (by placental compensatory effort), an attempt could be made to repeat these experiments of nature under controlled laboratory conditions.

If ovariectomized rats sustain pregnancy by placental compensation, then the combined procedures of ovariectomy plus placental dislocation must terminate pregnancy quickly and predictably. Placental dislocation at days sixteen or seventeen of gestation is readily performed. Animals so treated delivered 67 percent of the dislocated conceptuses in twenty-four hours.

However, the uterine volume in the placental dislocated horn decreases, and, therefore, after twenty-four hours the undelivered uterine contents are gradually reduced to a pulpy mass. To compensate for the volume reduction after placental dislocation, we shortened the horn at the ovarian end with a surgical suture. A group of twenty-two animals so treated delivered 81 percent of the dislocated fetuses in twenty-four hours. Important to note is that these ovariectomized plus placental dislocated rats did not begin to abort during or shortly after surgery. They started aborting on the average twelve hours following surgery, i.e. long after returning to their normal life, eating, drinking, and moving. This long latency period is evidence that abortion in

"ovariectomy syndrome" and labor

these animals is not due to surgical trauma, or manipulation, or the liberation of some oxytocic substance, but to the withdrawal of the luteoplacental support. Further evidence is offered by unilateral placental dislocation.

When only unilateral placental dislocation is carried out, in combination with bilateral ovariectomy, unilateral abortion occurs in the dislocated horn, while normal pregnancy is maintained in the intact horn (Fig. 4D). The dislocated horn delivered 86 percent of the fetuses while the intact horn only 6 percent. This finding is best explained by a local effect of placental progesterone, creating a functional asymmetry between the two uterine horns.[1]

The Electrophysiological Documentation of the Local Effect of the Placenta

In order to document further the local effect of the placenta and the resulting functional asymmetry of the uterus Dr. Hiroshi Takeda and I subjected these animals to electrophysiological studies. In these experiments the "suction electrode" method was used, a technique developed by heart physiologists.[19] This technique permits the rapid, in situ, placement of the recording electrodes under visual control and their equally rapid transposition from one uterine portion to another during the study. Thus, the technique permits the mapping up of the electric activity of the entire uterine horn, intact, or placental dislocated. In addition to electric activity the intrauterine pressure was also recorded, utilizing the extraovular microballoon method.

A series of experiments in twelve animals showed that forty-eight hours after unilateral placental dislocation there was a striking difference in the intrauterine pressure of the intact and placental dislocated uterine horns (Fig. 11A). The active pressure was regular, 50 mmHg on the average in the dislocated horn, while it was irregular and less than 10 mmHg in the intact horn. This is the mechanical evidence of a functional assymetry, resulting from the weakening of the progesterone block in the placental dislocated horn. The high and regular pressure observed in the placental dislocated horn can be triggered only by regular train discharges, in a propagating organ of considerable electric synchrony. The multiple suction electrode assembly showed this type of electric activity in the placental dislocated horn (Fig. 11A.)

prenatal life

Low and irregular pressure, observed in the intact horn, results from electric asynchrony. Pressure is low in the elastic yielding of the inactive regions. The suction electrode assembly documented (Fig. 11B, C, D) that the antiplacental region of the intact horn has distinct, but irregular, electric activity. However, as the electrodes were moved closer and closer to the placental bed the electric activity diminished (Fig. 11E) and practically vanished over the placental bed. This is evidence that functional placentae, in ovariectomized rats, exert a stronger block over nearby uterine regions than over distant regions, i.e. a local effect. Furthermore, the intact horn also revealed a

Figure VIII-11. The electric and mechanic activity of the uterus in ovariectomized rats after unilateral placental dislocation. Left: Schematic illustration of electrode and balloon positions; Right: original tracings. (a) The difference in mechanic activity between the placental dislocated (X) and the intact horn (Y); and the synchronic, regular electric activity of the dislocated horn. (b) The difference in electric activity between the dislocated horn and the anti-placental portion of the intact horn. (c-d) The closer is the electrode to the cervix the more regular is electric activity in the intact horn. (e) The electric silence over the placental portion of the intact horn. (f) The oxytocin effect. Note that the mechanic activity of the intact horn is not improved by oxytocin. (h) Recovery from oxytocin effect. (i-j) The electric activity of the near placental and the placental portions, before and after oxytocin treatment. Note the lack of oxytocin effect on the placental bed. (Courtesy, A. Csapo and H. Takeda, 1967.)

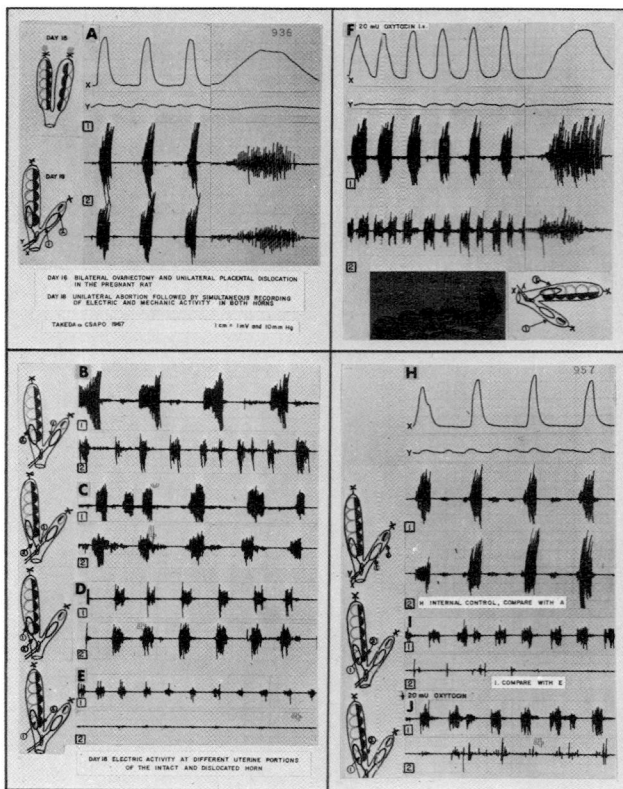

polarity in that the cervical end showed more advanced electric activity than the higher uterine regions (Fig. 11C, D). This explains the delivery of the first fetus in position, without detachment of the placentae at higher positions.

An intravenous injection of 20 mU oxytocin improved electric activity in the dislocated horn as well as in the antiplacental region of the intact horn (Fig. 11F). However, it only improved pressure in the dislocated horn. When the uterus recovered from the oxytocin effect (Fig. 11H, I) the internal control experiments showed that the character of activity is independent of the electrodes. Electric activity is determined by the position of the electrodes, i.e. their topographic relation to functional placentae. The suction electrodes also documented that oxytocin does not improve the electric activity of the placental bed markedly (Fig. 11J). This finding explains the unchanged pressure of the intact horn, during an oxytocin effect.

Evidently the oxytocin action is selective when the uterus is under placental control; the organ remains electrically and mechanically asymmetric. Under these conditions the pressure developed is determined by the ratio of the active/inactive uterine regions. Oxytocin improves activity in the already active and probably in borderline regions, but not in the more strongly blocked, inactive regions. This finding explains the recent observations[5] that around the seventh week of pregnancy there is no oxytocin response in human patients; that there is a distinct but limited response during early midtrimester pregnancy, when the luteoplacental shift is completed; and that the gradual evolution of the oxytocin response during second and third trimester pregnancy coincides with the changing ratio: nonplacental/placental uterine regions.[5]

Thus, the electrophysiological studies strengthened the interpretation of the physiological experiments and together with the steroid chemical data promoted the concepts that initiated this entire program. It seems that the rat by its luteoplacental compromise materially contributed to our understanding of myometrial regulation. The experimentally induced regulatory disturbances, threatening prenatal life, by their predictability and ready prevention offer useful experimental models for the study of similar disturbances that create prenatal hazards to the human fetus. Personally, I am most grateful to the rat for helping us to bridge the regulatory gap separating the rabbit and man.

prenatal life

The Initiation of Labor

From these experiments a simple concept emerges, readily explaining the mechanism that initiates labor in laboratory mammals and man. In appraising this concept it has to be recognized that "clinical labor," no matter how dramatic its onset (because of the subjective and objective symptoms of intense labor pains and rapid cervical progress), marks the end, rather than the beginning of a complex regulatory process. Clinical labor is preceded by a prolonged and gradual evolution in uterine activity. This clinically silent process has a widely different time table in different species. In some (for example, the rabbit) more sensitive and specific techniques than the recording of the intrauterine pressure can only document it reliably.

To cover a wide range we compared the time course of the evolution process in the two extreme species, the human and the rabbit, in technically similar trials. The recognition of the onset of the evolution process was promoted by oxytocin tests and in the rabbit by the recording of electric activity. Electric activity was also recorded in a third species, the rat, a suitable intermediate. On a unified gestational timetable, consisting of ten units, the evolution process manifested before delivery by eight time units in the human, 4 units in the rat, and one unit in the rabbit. This finding brings out a basic law, namely that the evolution process begins when the systemic progesterone of the corpus luteum is gradually withdrawn. Thus, it is the evolution process, not clinical labor, which is triggered by a regression of the corpus luteum.

This withdrawal of corpus luteum progesterone does not invariably manifest in decreasing peripheral plasma progesterone levels. Placental compensation may sustain or even increase the plasma progesterone levels demanded by the developmental stage of the fetus. Since in different species the fetal development is different at the time of the partial or more complete luteoplacental shift, the extent and duration of placental compensation are also different. As a result, uterine activity, triggering labor, evolves at different rates in different species, reflecting a compromise between luteal and placental contribution to the progesterone support of the myometrial block.

If the placental contribution to the block is minimal, as is the case, for example, in rabbits, luteal withdrawal dominates the timing of delivery. However, even under this extreme condition progesterone is only one regulatory parameter. Increasing uterine volume is another, promoting the evolution process. The volume effect in its terminal phase is facilitated by an oxytocin action. However, the volume effect is prolonged and lasting, while the oxytocin effect is short-lived and transient. This mechanism for the initiation of labor, dominated by a "true," systemic progesterone withdrawal, promotes rapid delivery.

A slightly modified mechanism, controlled by placental progesterone, is seen in patients when the evolution of uterine activity progresses in spite of increasing peripheral plasma progesterone levels. The rate of evolution under these conditions is determined by the V/P ratio, i.e. by an increasing placental failure in effectively balancing the activity-promoting effect of uterine volume. Thus, the evolution of uterine activity reflects a relative placental compensatory "failure" in balancing a volume-dominated regulatory process.

These two mechanisms for the initiation of labor are fundamentally similar. In both, progesterone, uterine volume, and oxytocin play similar regulatory roles. Only the relative contributions of the different factors to the overall balance vary, and the anatomical location of the source of progesterone. The experiments in ovariectomized rats underline the significance of the luteoplacental shift, and that of the change from systemic to local progesterone support of the myometrial block. This change has various regulatory consequences, as yet insufficiently studied. However, a multidisciplinary approach, such as that employed in this study, has the potential of exposing them and establishing their significance.

Acknowledgments

The author is grateful to Walter Wiest, Hiroshi Takeda, Martti Pulkkinen, Jacques Sauvage, Sadao Ogata, Yoshiharu Abe, Philippe Leger, and Elise Csapo for their various contributions to this work. The assistance of medical student Howard Aylward, Jr., during the initial phase of the rat experiments is appreciated.

References

1. Csapo, A. I.: Defence mechanism of pregnancy. In *Progesterone and the Defence Mechanism of Pregnancy*. Ciba Foundation Study Group No. 9, ed.

G. E. W. Wolstenholme and M. P. Cammeron, Boston, Little, Brown, 1961, p. 3-27.
2. Csapo A. I.: Model experiments and clinical trials in the control of pregnancy and parturition. *Am J Obst Gynec, 85:* 359-379, 1963.
3. Csapo, A. I.: Smooth muscle as a contractile unit. *Physiol Rev, 42:* 7-33, 1962.
4. Csapo, A. I.: The termination of pregnancy by the intra-amniotic injection of hypertonic saline. *1966-1967 Year Book of Obstetrics and Gynecology,* ed. J. P. Greenhill, Chicago, Year Book Medical Publishers, 1966, pp. 126-163.
5. Csapo, A. I. and Wood, C.: The endocrine control of the initiation of labor in the human. *Recent Advances in Endocrinology,* ed. H. T. James, 8th ed.; London, Churchill, 1968.
6. Csapo, A. I. and Csapo, E.: (Unpublished data).
7. Csapo, A. I.: Progesterone. *Scientific American, 198:* 40-46, 1958.
8. Haterius, H. O.: Reduction of litter size and maintence of pregnancy in the oophorectomized rat: evidence concerning the endocrine role of the placenta. *Amer J Physiol, 114:* 399-406, 1936.
9. Zeiner, F. N.: Studies on the maintenance of pregnancy in the white rat. *Endocrinology, 33:* 239-249, 1943.
10. Alexander, D. P., Frazer, J. F. D., and Lee, J.: The effect of steroids on maintenance of pregnancy in the spayed rat. *J Physiol, 130:* 148-155, 1955.
11. Kroc, R. L., Steinetz, B. G. and Beach, V. L.: The effects of estrogens, progestogens and relaxin in pregnant and nonpregnant laboratory rodents. *Ann NY Acad Sci, 75:* 942-980, 1959.
12. Abe, Y. and Csapo A.: (Unpublished data).
13. Ogata, S. and Csapo, A.: (Unpublished data).
14. Csapo, A. I. and Sauvage, J.: The evolution of uterine activity during human pregnancy. *Acta Obst et Gynec Scandinav, 47:* 181-212, 1968.
15. Fuchs, A. R.: Oxytocin and the onset of labour in rabbits. *J Endocrin, 30:* 217-224, 1964.
16. Caldeyro-Barcia, R.: Regulation of myometrial activity in pregnancy. In *Muscle,* ed. W. M. Paul, E. E. Daniel, C. M. Kay, and G. Monckton, Oxford, Pergamon Press, 1964, pp. 317-345.
17. Wiest, W. G.: Estimation of progesterone in biological tissues and fluids from pregnant women by double isotope derivative assay. *Steroids, 10:* 279-290, 1967.
18. Wiest, W. G., Kidwell, W. R., and Balogh, K., Jr.: Progesterone catabolism in the rat ovary: a regulatory mechanism for progestational potency during pregnancy. *Endocrinology, 82:* 844-859, 1968.
19. Hoffman, B. F., Cranefield, P. F., Lepeschkin, E., Surawicz, B., and Herrlich, H. C.: Comparison of cardiac monophasic action potentials recorded by intracellular and suction electrodes. *Am J Physiol, 196:* 1297-1301, 1959.

chapter IX

Studies on the Human Fetus

Carl Wood, M.D.

The accessibility of the human fetus to investigation has been established. The ability to collect blood from and attach electrodes to the fetal scalp during labor has enabled obstetricians to obtain new information. While present knowledge is limited both in extent and application, the techniques used to study the fetus can be expected to improve and eventually produce a marked change in the practice of obstetrics.

Studies on Fetal Scalp Blood

The technique of fetal scalp blood collection during labor was developed by Saling.[1,2] By using the small volume of blood collected from the scalp it has been possible to measure pH, pCO_2 pO_2, base, glucose, phosphate, sodium, potassium, chloride, haemoglobin, haematocrit, and insulin. The feasibility and accuracy of measuring substances in the fetal blood is determined by the availability of ultramicro methods of biochemical measurement and the ability to exclude contaminants entering fetal blood at the time of collection. The levels of scalp blood substances may also be influenced by variation of scalp blood flow. The reliability of measurements can be ascertained to some extent by the following procedures: checking the ultramicromethod against the method used for larger blood volumes; repeating estimation on blood collected from one scalp incision; measuring the substance in blood collected

prenatal life

from two separate incisions; measuring the substance in blood collected during and between contractions; and repeating the estimation on blood collected after an interval of time. In some situations interpretation of measurements can be made only if the relation of the time of blood collection to intra-amniotic pressure and heart rate changes is known.

One of the difficulties in establishing normal blood values for the fetus is that the criteria of fetal normality are based mainly upon the condition of the baby at birth (Apgar score), its weight, and its behavior in the neonatal period. However, a baby may be normal at birth but have suffered a temporary disturbance during labor. This temporary disturbance may coincide with fetal blood collection. However, a baby may behave normally both at birth and in the neonatal period but have abnormal neurological signs at one year of age. Until knowledge of fetal physiology inceases and the assessment of the condition of the newborn is more accurate, some reservation about the establishment of normal fetal blood values will remain.

pH

Reduction of fetal blood pH is associated with a reduction of blood base, an increase of pCO_2, or both (Fig. 1).

Reduction of fetal base may result from lowering of maternal base, the fetus reflecting this change because of transplacental equilibration of blood base.[3,4] Fetal base may also decrease when maternal base levels are normal. Such change may result from excess tissue acid production by the fetus in association with tissue hypoxia. Tissue hypoxia may result from failure of maternal-fetal oxygen transport or tissue perfusion. It is also possible that fetal base may be reduced in the absence of tissue hypoxia.

Fetal pCO_2 may increase as a result of obstruction of CO_2 transport from the fetus to the mother or as a chemical result of decreasing pH (Henderson-Hasselbach equation). In the absence of obstruction to CO_2 transport from the fetus to the mother the latter mechanism is unlikely to be important, because the normal placenta may function as efficiently as the lung[5] and remove excess CO_2. Sometimes fetal CO_2 rises as a result of changes in maternal pCO_2, e.g., heavy sedation, pulmonary disease.[6]

When CO_2 transport from the fetus is impaired, the fetus usually is suffering from lack of oxygen. If the lack of oxygen is severe enough to cause

the human fetus

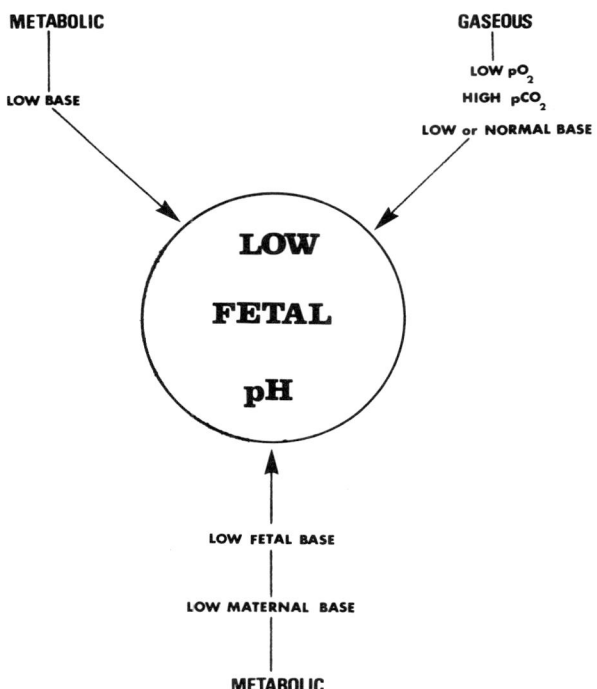

Figure IX-1   Causes of low fetal pH.

tissue hypoxia, tissue acid production will increase and base decrease. Thus, high pCO_2 is usually associated with low levels of base, both contributing to the fall in pH. Sometimes, however, fetal pCO_2 may be elevated while base levels remain normal. The occurrence and magnitude of decrease of fetal base in association with asphyxia depends on several factors. The rate and amount of increase in pCO_2 is important, because the higher the pCO_2 the more HCO_3^1 formed (Henderson-Hasselbach equation). Also, the quantity

prenatal life

of fixed acids produced in a tissue will depend not only on the levels of oxygen in the blood supplying the tissue but also on the adequacy of the circulation of that tissue and its pre-existing metabolic state. In addition, fetal base levels may depend upon the efficiency of the placenta in transporting acid from the fetus to the mother.[7,8] Rooth has calculated that the equivalent of 2.8 mEq/1 H^+ ions is transported across the placenta. If we were to assume a fetal placental blood flow of 0.51/min for a baby of 3.5 kg.,[9] then the normal placenta might transport as much as 90 mEq of acid per hour to the mother. A number of assumptions in these calculations limit their value,

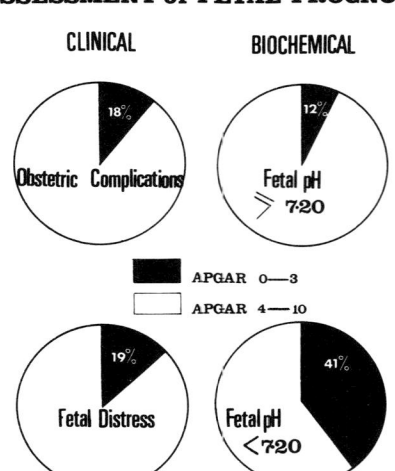

Figure IX-2. Clinical and biochemical analysis of 256 patients with obstetric complications known to be associated with increased fetal mortality; 152 of the 256 had clinical criteria of fetal distress (meconium in the liquor or FHR > 160 or < 120). Fetal blood pH was measured in the first stage of labor in all patients. Fetal blood pH measurement was superior to clinical criteria in assessing fetal prognosis ($p < 0.01$).

but they do provide a challenge for further investigation. It is not known whether excess production of acids by the fetus can ever alter maternal acid-base status. Since we have corrected our fetal base values for the degree of oxygen saturation of hemoglobin, mean fetal base deficit during labor, minus 3.3 mEq/1, has been less than the maternal base deficit, minus 5.7 mEq/1.

Most of the studies carried out on fetal blood have been concerned with measurement of fetal blood pH, as Saling has demonstrated that reduction of blood pH equilibrated at 40 mm Hg pCO_2 indicates a hazard to fetal life. Saling considers a fetal pH (equilibrated) of < 7.20 abnormal, but in determining the need for obstetric intervention he also takes into consideration the rate of decrease of fetal pH and the stage of labor. Beard et al.[10] have shown that where babies are born in poor condition, the fetal scalp blood pH taken within thirty minutes of birth is usually low. The closer to delivery that one collects fetal blood, the better correlation one can expect with the Apgar score, as James[11] has shown that cord blood pH values correlate with the clinical condition of the newborn. However, the obstetrician is more concerned with the value of fetal blood pH measurement in the first stage of labor. At this time the correlation of fetal blood pH with Apgar score is less dramatic, but it is significant. If the fetal pH is < 7.20, the baby has a 41 percent chance of being born with an Apgar score between 0 and 3, while if the fetal pH is > 7.20 only 12 percent are born with an Apgar score between 0 and 3 (Fig. 2). Furthermore, fetal blood pH measurement is more accurate in determining fetal prognosis than are the clinical criteria of fetal distress (Fig. 2). Only one in five patients with clinical distress has a low fetal pH, and the patients with a normal fetal pH can usually be managed conservatively. Because of the small number of stillbirths in any one series, it is difficult to obtain a correlation between low fetal blood pH and the occurrence of fetal death. Nevertheless, in a total series of 350 cases we have recorded four fetal pH values of less than 7.10 within eight hours of fetal death in utero; three of the four fetuses were premature.

The interpretation of low fetal pH may be improved by considering the basis and the rate of fetal pH change.

Base

Any reduction of fetal pH associated with reduced base demands examination of maternal base levels. If maternal base is also decreased and fetal pO_2 and pCO_2 are normal in several blood samples, then the cause of the fetal pH and base change is probably maternal.

Maternal and fetal base levels correlate significantly, and furthermore, the administration of acid (NH_4Cl) or base ($NaHCO_3$, THAM) will change fetal as well as maternal base levels.[4] It is difficult to determine the speed of equilibration of base or acid across the placenta, because maternal infusion of base cannot be very rapid. Recently, we completely corrected a maternal

prenatal life

and fetal base deficit of 12 and 8 mEq/1, respectively, in one hour by infusing 170 mEq of sodium bicarbonate.

What is the clinical significance of decreased maternal and fetal base? Vedra[12, 13] and Rooth and Nilsson[7] have studied maternal and cord blood at delivery in vigorous babies and have concluded that the reduction of fetal base at this time is maternal in origin. Beard[10] has also emphasized the importance of the relationship between maternal and fetal base levels and has used the term △ base deficit to describe the difference between them. Babies born in good condition but with low pH have normal △ base deficit, while babies born in poor condition with a low pH have a large △ base deficit. This can be restated as when a baby is born in good condition with a low pH, usually the low fetal pH is reflecting only a reduced maternal base.

In contrast, Rooth[33] has emphasized the importance of maternal pH in assessing fetal well-being. He showed that reduction of maternal pH is associated with reduction of fetal pH and argued on purely chemical grounds that lowering of fetal pH will increase pCO_2 (Henderson-Hasselbach equation) and reduce the oxygen-carrying capacity of fetal blood (Bohr effect). The practical importance of his hypothesis is limited to some extent by our own findings that neither maternal pH nor maternal level of base deficit during the first stage of labor is related to the condition of the baby at birth (Fig. 3).

However, some caution is warranted in assuming that reduced fetal base in the presence of reduced maternal base is not of clinical significance. A mother with severe preeclampsia had a pH of 7.32, a pCO_2 of 29 mm Hg, and a standard bicarbonate of 16 mEq/1; and the fetal blood values were pH 7.13, pCO_2 48 mm Hg, and standard bicarbonate 14.4 mEq/1. The △ base deficit was normal (3 mEq/1). The mother was given THAM to correct the low maternal and fetal base, and within two hours the mother's acid-base status was corrected, but the fetal base had further decreased (pH 6.79, pCO_2 79 mm Hg and standard bicarbonate 7.6 mEq/1). The fetus died twenty minutes after this blood collection. Although one cannot be certain that the fetal condition was abnormal at the time of the first collection, an abnormal heart rate pattern and the subsequent death of the fetus suggests this. Clinical data and the heart rate patterns should be considered before accepting an aetiological link between maternal and fetal base decrease and also in interpreting △ base deficit.

the human fetus

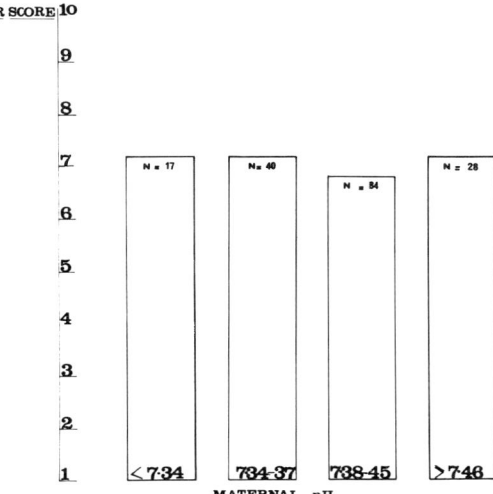

Figure IX-3. Relationship between Apgar score of the baby at two minutes and the maternal pH during the first stage of labor. The mean Apgar scores in the different groups were not significantly different.

Primary reduction of fetal base does occur and may be suspected when maternal acid-base status is normal, fetal pO_2 and pCO_2 is normal, but fetal base is reduced. However, confirmation of the diagnosis is difficult. Obstruction to gaseous interchange between the mother and fetus may be intermittent (e.g., during contractions) or temporary (e.g., tetanic contraction). If fetal blood is collected between the bouts of gaseous obstruction, then pCO_2 and pO_2 may be normal but fetal base be low as a result of excess acid production during the bout of hypoxia. To exclude this possibility as far as possible primary fetal base reduction should be diagnosed only when fetal blood gases are found to be normal both during and between contractions, and the measurements are repeated over a period of time.

We have studied three cases of primary fetal base reduction. A low fetal pH and a large fetal base deficit were present not only with normal maternal gases and acid-base values but also with fetal blood gases within the normal range. One patient had diabetes mellitus and the other two hypertension and proteinuria. In two of the three cases several blood samples were taken from the scalp, and these results were confirmed on cord blood. Two of the three babies had Apgar scores between 0 and 3.

Establishment of normal gas values in fetal scalp blood is limited by the factors previously discussed. At caesarean section Davey et al.[14] found

prenatal life

mean levels of scalp capillary blood pCO_2 and pO_2 between those of the umbilical artery and vein but closer to those in the artery.

pCO_2

In our series the mean fetal scalp pCO_2 was 42 mm Hg, the range of 2S.D. being 23.3-60.9. These results were taken from fetuses with a normal pH and heart rate who were subsequently born with an Apgar score of 6 to 10. Correlation between maternal and fetal pCO_2 has been demonstrated, and altering maternal pCO_2 by overbreathing or inhaling carbogen produces similar changes in fetal pCO_2.[6] The range of pCO_2 values found may have been increased as a result of the mother's overbreathing during labor, which decreases with maternal and fetal pCO_2. The effect of overbreathing on the fetus is controversial.[15-19] Variation of the conditions under which experiments are carried out and the techniques of blood gas measurement may account for the different results. Excessive overbreathing under general anesthesia has been shown to reduce the fetal oxygen tension in cord blood, possibly as a result of placental vasoconstriction.[15, 16] We have overbreathed six conscious patients during labor to maternal pCO_2 levels of fifteen to twenty mm Hg, and in only one did the fetal scalp capillary blood pO_2 fall significantly. Also, we have studied maternal pCO_2 and fetal scalp capillary blood pO_2 levels in eighty-seven patients during labor—overbreathing during contractions was frequent, as this technique is taught during antenatal preparation. No relationship was found between maternal pCO_2 and fetal pO_2 levels, and six patients with a maternal pCO_2 of < 20 mm Hg had normal fetal pO_2 levels (mean 24 mm Hg).

Increase of fetal pCO_2 beyond 60 mm Hg usually occurs with a low pH and a low level of base. However, a large increase of pCO_2 may occur with a low pH but normal levels of fetal base, e.g., patient E. R. had severe preeclampsia at thirty-eight weeks gestation and gave birth to a baby in poor condition (Apgar 1) that had some hypertonicity at six months of age. When the cervix was 6 cm dilated the fetal pH was 7.05, pCO_2 92 mm Hg, and the base deficit 6.4 mEq/1. The maternal pH was 7.40 and base deficit was 5.5 mEq/1—again △ base deficit was normal. Delivery occurred two hours later when the base deficit in umbilical venous blood had increased to 11.0 mEq/1. The case demonstrates the need to look at all acid-base parameters. During labor both the fetal base level and △ base deficit were normal, but the pH and pCO_2 were not.

pO_2

During the first stage of labor we found a mean scalp blood pO_2 of 22.2 mm Hg, the range being 12.8 to 31.6 (2 S.D.) These results were obtained in fetuses with pH > 7.20 which were subsequently born with an Apgar score between 6 and 10.

Controversy still exists as to whether increased oxygen inhalation by the mother increases fetal oxygenation.[20-24] Measurements on maternal and fetal capillary blood samples in normal and abnormal obstetric patients have demonstrated that maternal inhalation of 100 percent oxygen by mask for twenty minutes increases fetal blood pO_2[24] (Figs. 4, 5). In nineteen experi-

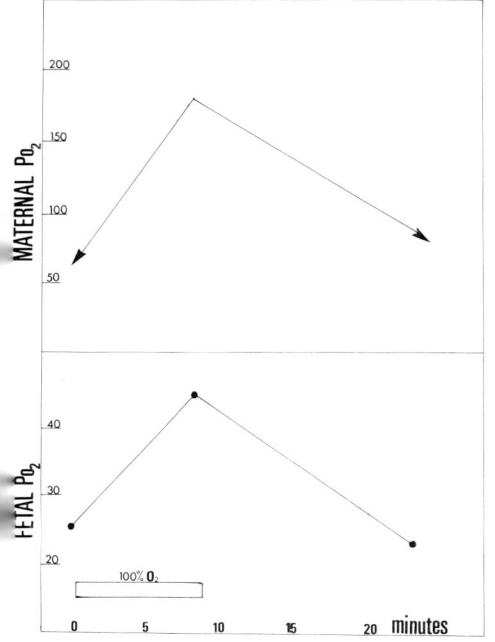

Figure IX-4. Maternal inhalation of 100 percent oxygen. Fetal pO_2 increased 19 mm Hg after maternal inhalation of 100 percent oxygen for eight minutes. Fetal pO_2 had returned to the baseline value fifteen minutes after withdrawal of the 100 percent oxygen.

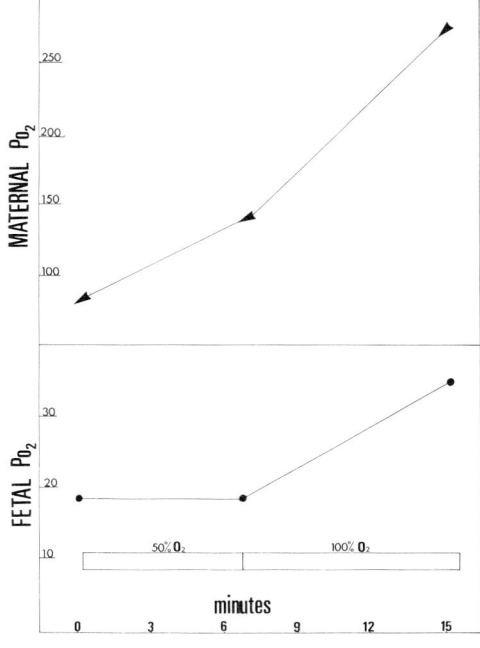

Figure IX-5. Comparison of the effect of 50 percent and 100 percent oxygen upon the level of fetal blood pO_2. In this patient 50 percent oxygen did not increase fetal pO_2. In contrast, 100 percent oxygen increased fetal pO_2 from 18 to 36 mm Hg.

prenatal life

ments the mean increase in fetal blood pO_2 was 14 mm Hg, which represents a 60 percent increase above the baseline level. The confidence limits of the techniques and the variation in baseline level were taken into consideration. In four patients 100 percent oxygen inhalation was continued for one hour. An increase of fetal pO_2 was maintained for the whole hour, although base deficit increased in one of the four fetuses. In situations where fetal hypoxia is present or imminent (e.g., at caesarean section), the inhalation of 100 percent oxygen, at least for short periods of time, may benefit the fetus.

When the mother breathed 10 percent oxygen for eight to thirteen minutes fetal pO_2 decreased 50 percent while fetal pH and base did not change.[24] Comparison of the fetal pO_2 levels in this experiment (12-13 mm Hg) with the fetal pO_2 in some abnormal obstetric situations is of interest. Fetal pO_2 during periods of slowing of the heart rate is usually higher than that during maternal inhalation of 10 percent oxygen. (Fig. 12). Also, the duration of experimental fetal hypoxia was longer. Nor during the maternal inhalation of 10 percent oxygen did the fetal pH or base levels change. In the past the importance of fetal pO_2 levels may have been overemphasized. Our own fetal pO_2 levels do not correlate well with the condition of the newborn. This may result partly from the fact that changes in fetal blood pO_2 are intermittent or transitory. However, circulatory changes in the fetus are also important. Vascular shunts or reduced blood flow in segments of the fetal circulation may cause local tissue hypoxia, which is sufficient to reduce fetal pH and base, while pO_2 in major vessels or the scalp remains normal. Furthermore, the susceptibility of the fetus to hypoxia may vary. In animals the resistance to hypoxia depends upon fetal maturity and cardiac glycogen stores.[25] It is possible, therefore, that temporary hypoxia, which is difficult to detect in the human being during labor, may be more deleterious to a susceptible fetus than is prolonged hypoxia to a healthy fetus. In only one of sixteen fetuses born in poor condition (Apgar 0 to 3) was fetal pO_2 low during labor ($<$ 14 mm Hg), yet eleven of these sixteen fetuses showed an abnormal heart rate or low pH.

Glucose

Carbohydrate metabolism is important to both the fetus and the neonate.[25, 26] There is both experimental and clinical evidence that the extent of cerebral cortical damage in fetal hypoxia is in part related to the availability of glucose, large amounts of glucose and alkali having a protective effect.[27, 28] Hypoxia is itself a factor in altering blood glucose levels. In acute asphyxia blood glucose may increase, while in prolonged asphyxia it may decrease as a result of depletion of hepatic glycogen stores. The intensive studies of carbohydrate metabolism in the neonate and animal fetus led us to carry out a number of studies on human fetal blood glucose levels.

During labor maternal and fetal glucose concentrations are related with a positive gradient from mother to fetus.[29] At birth there is a small increase of both maternal and fetal blood glucose levels. Following rapid injection of glucose in parturient mothers, comparable elevations of maternal and fetal glucose occur, and both levels return to normal within two hours[29] (Fig. 6). Glucose transfer across the placenta is thought to involve active transport,[30, 31] and in the human at least the transport of glucose may be oxygen dependent.[32] Following the maternal inhalation of 10 percent oxygen, maternal glucose concentration increased 35 mg percent, but fetal glucose levels did not change. A comparable elevation of maternal glucose, 32 mg percent, can be produced by the intravenous injection of 10 G glucose, and this is accompanied by a similar increase of fetal blood glucose, 33 mg percent. The most likely explanation of the failure of fetal blood glucose to increase after the maternal inhalation of 10 percent oxygen is that glucose transport across the placenta is oxygen dependent.[32]

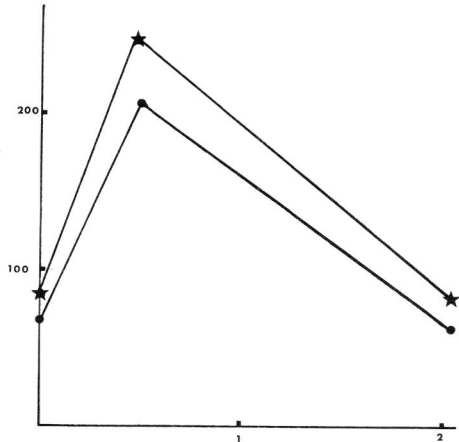

Figure IX-6. Mean maternal (*) and fetal (o) blood glucose levels following rapid intravenous injection of 50 g. of glucose in six mothers.

prenatal life

During labor low levels of fetal blood glucose have been observed. A blood concentration of more than two standard deviations below the mean was defined as fetal hypoglycemia ($<$ 40 mg percent).[33] This occurred in association with four conditions: fetal growth retardation, preeclampsia, abruptio placenta, and maternal hypoglycemia. Two of the eight fetuses with hypoglycemia were stillborn.

Fetal Growth Retardation

In three patients there was fetal growth retardation. One of these had severe preeclampsia and one had prolonged pregnancy, but the remaining patient had no other clinical abnormality. In two of the patients maternal urinary excretion of estriol was abnormal, and fetal blood pH was low (7.13 and 7.16). One fetus died before birth (glucose 38 mg percent), and the other two were born with Apgar scores of 4 and 6, the glucose levels being 30 and 14 mg percent, respectively. The latter, which had the lowest glucose in the series, was delivered by elective caesarean section at thirty-eight weeks and was clinically dysmature; subcutaneous fat was absent, the skin was dry and peeling, and it weighed only 1825 g.

Preeclampsia

In two patients preeclampsia was present, but in contrast to the previous group these babies had a birth weight consistent with their gestation.

Abruptio Placenta

The third situation in which fetal hypoglycemia occurred was in the only two patients in the series with accidental hemorrhage. Again birth weight was normal. One fetus was stillborn; the fetal blood pH was 6.60 and the glucose 20 mg percent just prior to death. In these three situations the maternal glucose level was normal.

Maternal Hypoglycemia

In one patient maternal hypoglycemia caused fetal hypoglycemia (20 mg percent). The mother, whose blood glucose in labor was 45 mg percent, had

the human fetus

fasting blood total reducing sugars of 63 mg percent in the last trimester and showed glycosuria with normal blood total reducing sugars on two occasions. During labor both maternal and fetal blood glucose increased by 30 mg percent in response to an intravenous load of 10 g glucose—the expected increase being 30-40 mg percent. The baby was born in a satisfactory condition, and its subsequent behavior was normal.

Of the six fetuses surviving birth one developed a subdural hematoma, but the remainder behaved normally. Blood glucose was measured in only one of the babies during the first twenty-four hours after birth—it was 24 mg percent.

The level of fetal blood glucose will depend upon the amount transferred across the placenta, the rate of storage and release of fetal glycogen, and the peripheral utilization of glucose. There was evidence of disturbance of placental function in some of our patients (lowered urinary estriol excretion, placental separation), which may have contributed to the fetal hypoglycemia. Acute asphyxia produces a rapid fall in cardiac and liver carbohydrate in both fetal and neonatal animals,[34, 35] and this mechanism may have contributed to the fetal hypoglycemia in the two patients with abruptio placenta.

Fetal blood glucose levels were not related to fetal blood pO_2. The rapid transfer of glucose from the mother to the fetus may obscure changes in fetal blood glucose resulting from hypoxia. Alternatively, fetal blood glucose levels may not change during asphyxia or change only when asphyxia is severe.

Romney has produced evidence that glucose administration during labor may prevent fetal metabolic acidosis.[36] We found no relationship between levels of fetal blood glucose and base deficit, and, furthermore, there was no consistent change in fetal base deficit after a maternal glucose load of 50 g.[33] If placental transfer of glucose is impaired, treatment of fetal hypoglycemia by a maternal glucose load may be ineffectual. However, transport may not be the dominant factor determining fetal hypoglycemia, because in two patients this has been temporarily corrected by a maternal glucose load (Fig. 7). In practice, it would be best to treat mothers in whom fetal hypoglycemia is more likely, i.e., fetal growth retardation, preeclampsia, and abruptio placenta, as the diagnosis in an individual patient is impracticable. By treatment it is hoped to replenish fetal glycogen stores. There is some evidence that this can be achieved in the neonate.[37]

prenatal life

In an attempt to understand other factors controlling fetal blood glucose levels during labor we have measured maternal and fetal blood insulin levels following a maternal glucose load and also after maternal injection of insulin.[38] Following an intravenous injection of 50 g glucose, maternal insulin levels and maternal and fetal glucose concentrations increase, but fetal blood insulin does not change. After a maternal injection of 0.1 U insulin, kg maternal insulin levels increase, while both maternal and fetal blood glucose decrease. Again fetal blood insulin levels do not change. It was apparent that the fetal pancreas responded little, if at all, to changes in fetal blood glucose levels. In order to determine whether the fetus can alter its blood glucose level independent of the mother, an injection of glucagon was made directly into the fetal scalp during labor.[33] Following the injection of 1 mg glucagon, fetal blood glucose increased 19 mg percent, while maternal blood glucose increased by only 3 mg percent. As the human placenta has low glycogen

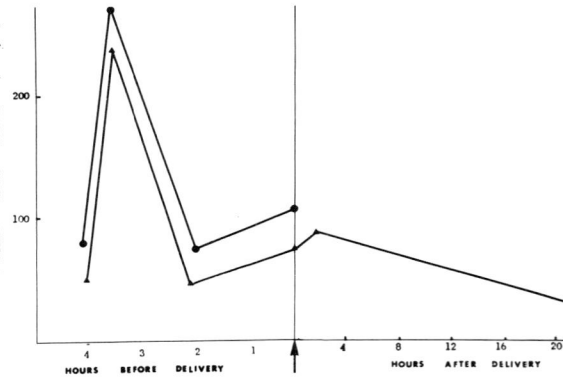

Figure IX-7. Effect of injection of 50 g. of glucose in a patient with hypertension and proteinuria. Fetal blood glucose level (Δ) was low (42 mg percent) but rose rapidly to 212 mg percent after the maternal dextrose load. Neonatal blood glucose fell to 23 mg percent twenty-four hours after birth. Maternal blood glucose (o).

stores at term,[39] glucagon probably raised the fetal glucose level by mobilizing fetal glycogen.

Ketones

The recognized tendency of pregnant patients to develop ketonemia and ketonuria has been confirmed by an investigation of maternal blood and urine ketone concentrations.[40] The significance of ketonaemia remains in doubt. There was no tendency for those patients with the highest ketone levels to have either a maternal or a fetal acidosis. Nor were blood glucose levels low in patients with ketosis. However, because excess ketones decrease peripheral utilization of glucose, it is sensible to continue the current practice of treating ketosis when it is discovered in early labor. A reliable method for doing this is the intravenous injection of 50 g glucose.

There is a correlation between ketone concentrations in maternal and cord blood. One interesting finding was a relative increase of fetal cord blood ketones in five patients with preeclampsia. In one of these the fetal level was greater than the maternal. It is possible that either fetus or placenta was breaking down fat, which contrasts with the normally exclusive use of carbohydrate by the fetus.

Studies with Fetal Scalp Electrodes
Measurement of Tissue pO_2 of Human Fetal Scalp

Collection of fetal scalp blood enables measurement of fetal blood pO_2, but the information obtained is relevant only to the time of collection. Measurement of fetal scalp tissue pO_2 may be a way of monitoring fetal oxygenation continuously. We have developed a membrane-covered flush-type oxygen electrode for this purpose[41] (Fig. 8). Movement artefacts were reduced by the membrane coating and by the matching of the electrode diameter to the mean intercapillary distance, the electrode then measuring mean tissue pO_2.

Although the electrode was not considered accurately quantitative (on theoretical grounds), the tissue pO_2 measured was always less than that of scalp capillary blood. In five patients in whom both tissue and capillary blood pO_2 were measured, the mean blood pO_2 was 20 mm Hg, and the mean tissue pO_2 was 9 mm Hg.

By using this electrode we have demonstrated that maternal inhalation of 50 percent oxygen may increase fetal scalp tissue pO_2 (Fig. 9). In two patients with slowing of the fetal heart rate during contractions, scalp tissue pO_2 decreased at the time of heart rate change (Fig. 10). After birth, scalp tissue pO_2 may mirror the condition of the newborn. When the baby was born in good condition, the scalp tissue pO_2 increased rapidly after birth, while in one baby born in poor condition scalp tissue pO_2 remained at fetal levels until the baby was intubated and oxygenated. Because of the difficulty

prenatal life

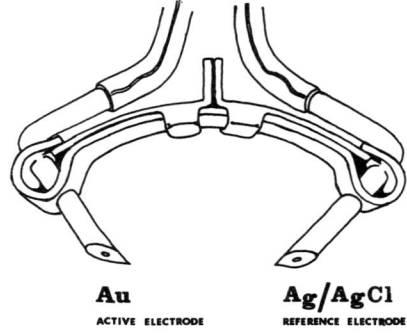

Au
ACTIVE ELECTRODE

Ag/AgCl
REFERENCE ELECTRODE

Figure IX-8. The scalp tissue pO_2 electrode. Both active and reference electrodes are mounted in a 14 mm wound clip. The active electrode is a 60μ gold wire coated with dichloro dimethyl hydrosilicate. The gold wire is coated with araldite and threaded into a nylon tube inside the needle protruding from the clip.

in making this electrode and its limited accuracy of measurement, it has no use at present in clinical obstetrics.

Measurement of Temperature in Human Fetal Scalp

An electrode has been developed for the continuous measurement of temperature in the human fetal scalp[42] (Fig. 11). Fetal temperature has been measured in the scalp and rectum.[43] When compared to the maternal rectal, vaginal, or intrauterine temperature, the fetal temperature is 0.3 to 1.9°C higher. In one patient the fetal temperature increased independently of the mother. This occurred during the development of fetal tachycardia and may be related to failure of the fetus to lose heat across the placenta or to increased heat production by the fetus.

Figure IX-9. Rise of fetal scalp tissue pO_2 during maternal inhalation of 50 percent oxygen. Fetal scalp capillary blood pO_2 was also measured and increased from 14 to 26 mm Hg after ten minutes.

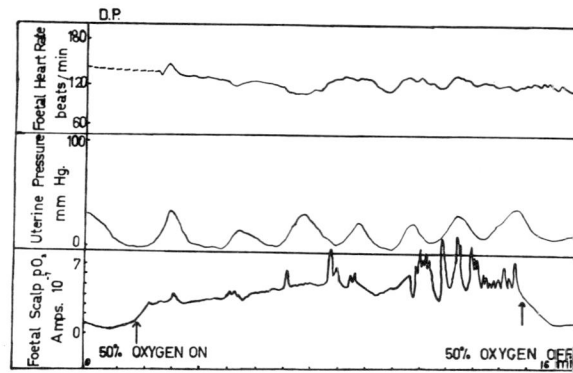

the human fetus

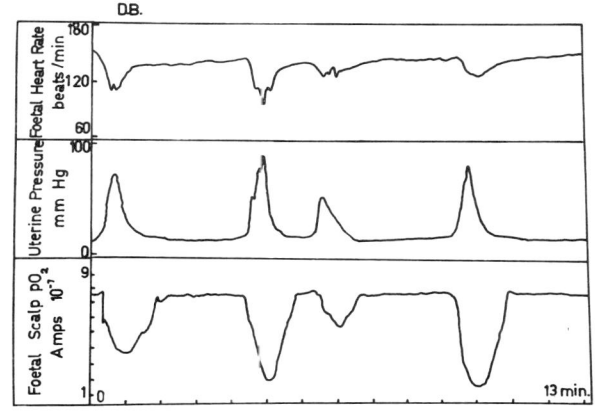

Figure IX-10. Slowing of the fetal heart rate with coincident decrease of fetal scalp tissue pO_2.

Fetal Heart Rate

By studying fetal heart rate patterns during labor Hon has been able to distinguish changes of heart rate that correlate with the condition of the baby at birth.[44] By scoring the changes in fetal heart rate according to their nature and severity in the thirty minutes before birth he found a close correlation with Apgar score. To obtain fetal heart rate (FHR) Hon has used a scalp ECG electrode whose impulse is fed into a beat-to-beat ratemeter, the heart rate then being displayed on a recorder.[45]

Figure IX-11. The scalp tissue temperature probe.

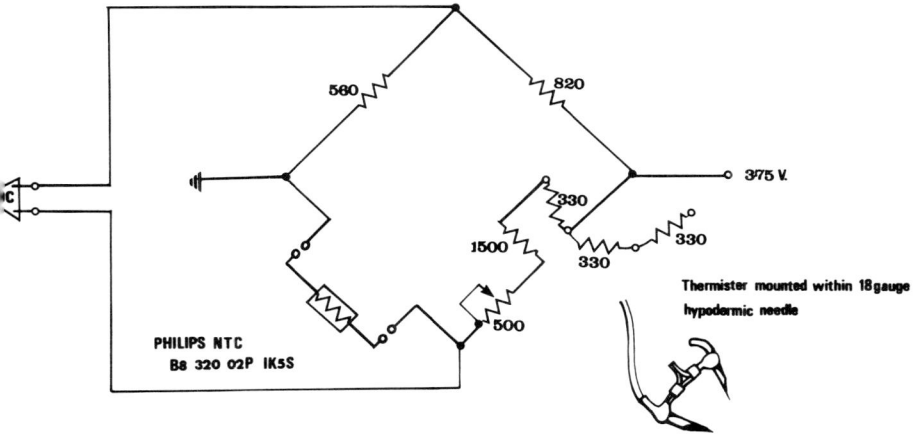

prenatal life

Fetal heart rate monitoring is a continuous process and therefore has the advantage of providing an early warning to the obstetrician of the presence of an unexpected fetal complication. However, it is expensive and requires skilled technical assistance. It is possible that by using ultrasonic techniques a cheap external method of recording heart rate may be developed.

One of the difficulties in the interpretation of fetal heart rate change is the lack of information concerning their physiological bases. Fetal tachycardia may be associated with an increase of fetal pCO_2 or decrease of fetal base.[4, 6]

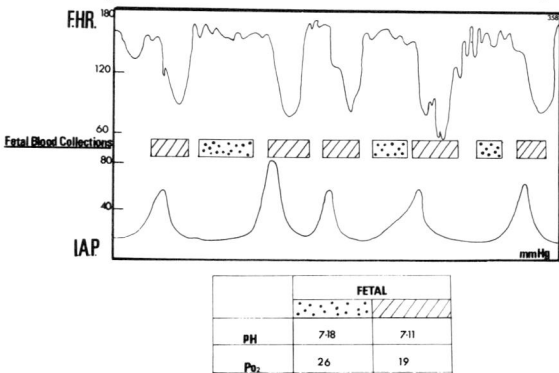

Figure IX-12. Change in fetal scalp blood measurements during fetal bradycardia. Fetal blood was collected during four bouts of fetal bradycardia and compared with that collected when FHR was normal. During the fetal bradycardia, fetal scalp blood pH and pO_2 decreased and pCO_2 increased slightly.

Unless it is accompanied by heart rate changes during contractions, it does not often signify a fetal hazard.[44, 46]

Hon has suggested that the slowing of the heart rate early or during a contraction usually is due to compression of the umbilical cord. Evidence for this is its more frequent association with the presence of the cord around the fetal neck, its occurrence in association with cord prolapse, and the artificial production of the same pattern at caesarean section by cord compression. Using fetal blood collection in association with fetal heart rate monitoring, Hon has shown that the fetal pH decreased over a period of time when the fetal heart rate slowed with contractions. We have collected fetal blood between contractions when the FHR was normal and during contractions when the FHR was slow. Usually there is no change in fetal gas and base during contractions, but in five cases with marked slowing during contractions ($>$ 60 beats/min), gaseous obstruction was evidenced

the human fetus

by the decrease of pO₂ and the increase of pCO₂. In one patient blood was collected during four periods of bradycardia and the pooled sample measured (Fig. 12). Additional evidence of gaseous obstruction during this FHR change was obtained with the scalp tissue pO₂ electrode (see Fig. 10), the value decreasing during the slowing of the heart rate. The change in fetal gases is not necessarily the cause of the fetal heart rate change, as lower fetal blood gas values are sometimes found in the presence of a normal fetal heart rate.

Hon has also described a pattern of heart rate change associated with placental insufficiency.[45] The heart rate slows at the end of a contraction, and this is often linked with loss of the beat-to-beat irregularity that normally occurs and also with tachycardia between contractions. Babies born with this pattern of FHR change often have low Apgar scores,[44, 46] but the basis of the pattern is not certain. When fetal blood is collected during the FHR change and between contractions, fetal gaseous and acid-base change sometimes occurs.

Clinical Assessment of Fetal Diagnostic Methods

In order to assess the clinical value of fetal diagnostic tests carried out during labor we analysed the records of forty-seven babies born with an Apgar score of 0 to 3 at two minutes. The tests included the recording of fetal heart rate and the measurement of fetal blood pH, pCO₂, pO₂, base deficit and glucose. During the second stage of labor fetal blood pH (< 7.20) and FHR were usually abnormal. However, during the first stage these changes were less common. When it was possible to analyze the time sequence of the onset of FHR and fetal pH changes, a significant FHR change nearly always preceded a significant fall of blood pH. Measurements of fetal scalp blood pO₂ and pCO₂ were less useful as prognostic tests than either FHR or fetal blood pH. Nutritional and metabolic factors may also be important in determining fetal condition. This was suggested by the frequency of fetal growth retardation (7/47) and the occurrence of fetal hypoglycemia in two patients.

Fetal diagnostic medicine is now established. However, its future must depend not only upon improving present diagnostic methods, but also upon understanding the importance and interrelationships of altered fetal metabolic, nutritional, circulatory, acid-base, and gaseous function.

Acknowledgments

This is mainly a review of the work carried out by the following people: Dr. J. W. Newman, J. F. Leeton, P. J. Paterson, A. Walker, D. Walker, Mrs. L. Wallace, Miss J. Hammond, and Miss L. McKinnon in our department and Dr. P. Taft and Miss J. Sheath of the Diabetic and Metabolic Unit, Alfred Hospital, and the Department of Medicine. We also wish to thank Dr. J. Owen of the Alfred Hospital for his advice. We are grateful for financial support from the Felton Bequest.

References

1. Saling, E.: A new method for examination of the child during labor: Introduction, technic and principles. *Arch Gynaek, 197:* 108, 1962.
2. ———: A new method of safeguarding the life of the foetus before and during labour. *J Int Fed Gynaec Obstet, 3:* 101, 1965.
3. Rooth, G.: Early detection and prevention of foetal acidosis. *Lancet, 1:* 290, 1964.
4. Newman, W., Mitchell, P., and Wood, C.: Fetal acid-base status. II: Relationship between maternal and fetal blood bicarbonate concentrations. *Amer J Obstet Gynec, 97:* 52, 1967.
5. Adamsons, K.: Birth defects. *Symposium on the Placenta, 1:* 27, 1965.
6. Newman, W., Braid, D., and Wood, C.: Fetal acid-base status. I: Relationship between Maternal and Fetal pCO_2. *Amer J Obstet Gynec, 97:* 43, 1967.
7. Rooth, G., and Nilsson, I.: Studies on foetal and maternal metabolic acidosis. *Clin Sci, 26:* 121, 1964.
8. Rooth, G.: In *Development of the Lung*, Ciba Symposium, ed. A. V. S. De Reuck and R. Porter, London, Churchill, 1967.
9. Assali, N. S.: Some aspects of fetal life in utero and the changes at birth. *Amer J Obstet Gynec, 97:* 324, 1967.
10. Beard, R. W., Morris, E. D., and Clayton, S. G.: pH of fetal capillary blood as an indicator of the condition of the fetus. *J Obstet Gynaec Brit Comm, 74:* 812, 1967.
11. James, L. S.: Acidosis of the newborn and its relation to birth asphyxia. *Proceedings IX. Congress of Pediatrics*, Montreal, 1959.
12. Vedra, B.: Acidosis and anaerobiosis in full term infants. *Acta Paediat, 48:* 60, 1959.
13. ———: "Partial anaerobiosis" in the human fetus. Oxygen content and arteriovenous lactate differences in umbilical cord blood. *Amer J Obstet Gynec, 86:* 1088, 1963.

14. Davey, D. A., du Toit, H. J., Farrell, A. G. W., Rorke, M., and Tresize, R.: *Proceedings of the Fifth World Congress of Obstetrics and Gynaecology*, ed. C. Wood and W. A. W. Walters, Sydney, Butterworths, 1967.
15. Moya, F., Morishima, H. O., Shnider, S. M., and James, L. S.: Influence of maternal hyperventilation on the newborn infant. *Amer J Obstet Gynec, 91:* 76, 1965.
16. Morishima, H. O., Daniel, S. S., Adamsons, K., Jr., and James, L. S.: Effects of positive pressure ventilation of the mother upon the acid-base state of the fetus. *Amer J Obstet Gynec, 93:* 269, 1965.
17. Selwyn Crawford, J.: Maternal hyperventilation and the foetus. *Lancet, 1:* 430, 1966.
18. Motoyama, E. K., Rivard, G., Acheson, F., and Cook, C. D.: Adverse effect of maternal hyperventilation on the foetus. *Lancet, 1:* 286, 1966.
19. Coleman, A. J.: Absence of harmful effect of maternal hypocapnia in babies delivered at caesarian section. *Lancet, 1:* 813, 1967.
20. Prystowsky, H.: Fetal blood studies. *Amer J Obstet Gynec, 78:* 483, 1959.
21. Oliver, T. K., Jr., Demis, J. A., and Bates, G. D.: Serial blood-gas tensions and acid-base balance during the first hour of life in human infants. *Acta Paediat, 50:* 346, 1961.
22. James, L. S.: Adaptation to extra-uterine life. Report of Thirty-First Ross Conference on Pediatric Research. Columbus, Ross Laboratories, 1959.
23. Caldeyro-Barcia, R.: Symposium on the effects of labor and delivery on fetus and newborn, Oxford, Pergamon, in press.
24. Newman, W., McKinnon, L., Phillips, L., Paterson, P., and Wood, C.: Oxygen transfer from mother to fetus during labor. *Amer J Obstet Gynec, 99:* 67, 1967.
25. Shelley, H. J., and Nelligan, C. A.: Neonatal hypoglycaemia. *Brit Med Bull, 22:* 34, 1966.
26. Shelley, H. J.: Glycogen reserves and their changes at birth and in anoxia. *Brit Med Bull, 17:* 137, 1961.
27. Dawes, G. S., Mott, J. C., Shelley, H. J., and Stafford, J.: The prolongation of survival time in asphyxiated immature foetal lambs. *J Physiol* (London), *168:* 43, 1963.
28. Dawes, G. S., Hibbard, E. and Windle, W. F.: The effect of alkali and glucose infusion on permanent brain damage in rhesus monkeys asphyxiated at birth. *J. Pediat, 65:* 801, 1964.
29. Paterson, P., Phillips, L., and Wood, C.: *Amer J Obstet Gynec.* In press.
30. Ely, P. A.: The placental transfer of hexoses and polyols in the guinea-pig, as shown by umbilical perfusion of the placenta. *J Physiol* (London), *184:* 255, 1966.
31. Huggett, A. St. G.: Carbohydrate metabolism in the placenta and foetus. *Brit Med Bull, 17:* 122, 1966.
32. Lumley, J., and Wood, C. (Unpublished manuscript)

33. Phillips, L., Lumley, J., Paterson, P., and Wood, C.: Fetal hypoglycemia. *Amer J Obstet Gynec, 102:* 371, 1968.
34. Dawes, G. S., Mott, J. C., and Shelley, H. J.: The importance of cardiac glycogen for the maintenance of life in foetal lambs and new-born animals during anoxia. *J Physiol* (London), *146:* 516, 1959.
35. Stafford, A., and Weatherall, J. A. C.: The survival of young rats in nitrogen. *J Physiol* (London), *153:* 457, 1960.
36. Romney, S. L., and Gabel, P. V.: Maternal glucose in the management of fetal distress. *Amer J Obstet Gynec, 96:* 698, 1966.
37. Scott, J. M.: Histological observations on glycogen reserves of the foetus and newborn infant. *Arch Dis Child, 40:* 317, 1965.
38. Paterson, P., Taft, P., Phillips, L., and Wood, C.: Study of maternal and fetal blood insulin levels during labor. (Unpublished manuscript)
39. Villee, C. A.: Regulation of blood glucose in the human fetus. *J Appl Physiol, 5:* 437, 1953.
40. Paterson, P., Sheath, J., Taft, P., and Wood, C.: Maternal and foetal ketone concentrations in plasma and urine. *Lancet, 1:* 862, 1967.
41. Walker, A., Phillips, L., Powe, L., and Wood, C.: A new instrument for the measurement of tissue pO_2 of human fetal scalp. *Amer J Obstet Gynec, 100:* 63, 1968.
42. Walker, D.: An electrode to continuously measure human fetal scalp temperature. (Unpublished manuscript)
43. Wood, C., and Beard, R. W.: Temperature of the human foetus. *J Obstet Gynaec Brit Comm, 71:* 768, 1964.
44. Hon, E. H.: *Proceedings of Fifth World Congress of Obstetrics and Gynaecology,* ed. C. Wood and W. A. W. Walters, Sydney, Butterworths, 1967.
45. ———: Observations on "pathologic" fetal bradycardia. *Amer J Obstet Gynec, 77:* 1084, 1959.
46. Wood, C., Ferguson, R., Leeton, J., Newman, W., and Walker, A.: Fetal heart rate and acid-base status in the assessment of fetal hypoxia. *Amer J Obstet Gynec, 98:* 62, 1967.

chapter X

Exogenous Factors Affecting Labor

Emanuel A. Friedman, M.D., Sc.D.

Assertions and contradictions punctuate currently available data pertaining to the pharmacological action of many drugs on the gravid uterus. Because of this, obstetrical care today is tempered with middle-of-the-road principles, which stress individualization and due consideration of any inherent risks pertaining to the drugs used. The latter, particularly, concern potential effects on fetal function and on the course of labor. With regard to labor there is such marked variability and acknowledged subjectivity of evaluation that reports of divergent actions of drugs are prevalent. Many studies indicate opposing responses with regard to inhibition or stimulation of uterine contractions. In vivo investigations are often at a variance with those carried out in vitro, and animal studies sometimes differ from one another and from observations in humans. Species specificity might explain these discrepancies, but they are more likely due to inconsistencies in experimental conditions and control.

The physiology of myometrial contractility and the course of labor in the human parturient have been studied in many ways. While there has been much activity along these lines, investigation of individual variables has not consistently provided a true picture of the overall mechanistic complex. The "efficiency" of the machine with which we are dealing is not yet fully characterized. Diminished uterine contractility with shorter, less frequent, and less intense contractions is not necessarily accompanied by diminished pro-

prenatal life

gression in cervical dilatation or in descent of the fetus through the birth canal. No single factor measures overall performance in a manner that is suitably objective. Nevertheless, much basic information is known; perhaps the future will achieve total integration and rationalization of conflicting evidence.

My aim here is, of necessity, an incomplete review of the subject. There are enormous numbers of publications reporting on the many agents used during labor. I wish merely to present documented effects where these can be defined, and to delineate those controversial issues of clinical pertinence.

Analgesics and Sedatives

Morphine has been shown to have a variable effect. This variation has been explained, on a conjectural basis only, as the result of the uncovering of a deficiency of myometrial action by morphine. The same inconsistency is seen with meperidine. Discrepant experimental evidence notwithstanding, the overall clinical impression is one of diminished uterine activity and progress. Altered cervical dilatation-time pattern, a clinical measure of progression, is found where the patient is heavily sedated. This is especially evident when sedation is administered early in the first stage, prior to active dilatation. There is a difference in the response of the uterus to sedation in the considerably more sensitive latent phase, as contrasted to that in the active phase of the first stage. This difference may offer a partial explanation for some of the variable results encountered in experimental studies of contractility in labor. It is reasonable, for example, to expect major inhibition of uterine contractility when the narcotic agent is given early in the latent phase; but there is perhaps no effect at all during the active phase or second stage, unless the dosage is increased correspondingly.

Unverified isolated reports have indicated unpredictable responses among various narcotic-analgesics. Heroin is said to have no appreciable effect, while methadone diminishes uterine tone somewhat; labor appears to be shortened in uncontrolled clinical studies evaluating oxymorphone; and alphaprodine and phenazocine showed high degrees of variability when studied clinically on the basis of duration of labor.

Large doses of barbiturates may seriously inhibit uterine contractility and slow the progress of clinical labor. A dose-response relationship appears to exist for pentobarbital sodium and thiopental sodium. No effect is seen with minimal doses; diminished frequency and duration of contractions occur with larger doses; and complete cessation of uterine activity takes place with anesthetic doses. On the other hand, phenobarbital has no inhibiting effect on isolated human uterine muscle, nor does amobarbital on labor studied by external hysterography.

Tribromoethanol prolongs the interval between contractions, delays labor even in low doses, and may produce marked uterine atony. Paraldehyde is also capable of prolonging labor and decreasing uterine activity, particularly in amnesic dosages. The earlier it is given, the longer and more persistent its effect, and the more difficult the inertia produced is to overcome.

The narcotic antagonists, N-allyl normorphine and levallorphan, act upon the uterus in a variable manner similar to that of meperidine but to a lesser degree. There has been no counteraction to the depressive effect of sedative-analgesic drugs on the uterus.

Tranquilizers

Concerning the various currently used ataractic drugs, studies have appeared in the form of controlled objective analyses and clinical reports, most of which are testimonials or at best less than ideally designed experiments. Promazine action ranges to the extremes, from inhibition to stimulation. Unfortunately, objective investigations, utilizing intrauterine pressure recording devices, are just as inconsistent. Reported studies show no effect, in vitro depression of spontaneous activity in proportion to the concentration, and in vivo diminution in amplitude and frequency of contraction, together with arrested clinical progress in labor.

Chlorpromazine also is reported to prolong or shorten labor. Even intra-amniotic pressure recordings show variable effect. The recurring theme in relevant studies is the simultaneous administration of other sedative-analgesic medications that confuse the picture. One might interpret the results to indicate that the diminished amount of sedative necessary for pain relief augments contractions. Evaluation of the pertinent effect on uterine contractility is thereby thoroughly clouded.

Labors are shortened when promethazine is given, and the requirements for analgesic agents are acknowledged to be diminished. These labors

prenatal life

should be expected to be shorter than comparable labors in which more analgesics are administered. I have already mentioned how sensitive the uterus is to sedative-analgesic drugs, particularly during the early latent phase of the first stage. The foreshortening, therefore, is not necessarily due to any direct uterotonic action, but rather is more likely a secondary benefit resulting from the lesser amounts of narcotics required for pain relief. It is interesting to note that, objectively in physiologic studies, promethazine characteristically inhibits both amplitude and frequency of uterine contraction.

Similarly diverse results are obtained with perphenazine, with some clinical investigations showing no effect and others a slowing effect and still others an enhancing effect. As to prochlorperazine, most studies deny any real effect. One well-controlled, objective experiment shows delay of the patient's subjective awareness of uterine contraction, but no apparent direct effect upon uterine contractility.

In recent evaluation of chlordiazepoxide (Librium) no response is seen tokographically. At higher doses marked relaxation with diminished frequency or cessation of spontaneous contraction is reported. No change at low doses, increased tone and contractile incoordination at moderate doses, and diminished resting tonus and spontaneous contractility at high doses are also seen in a conflicting study using both internal and external tokography. Objectivity notwithstanding, therefore, inconsistent response may only reflect differences in conditions applied to the several experiments.

The antihistaminic, dimenhydrinate (Dramamine), shortens labor in uncontrolled clinical studies. Since no corrections are made for the differences that might be attributable to the analgesic potentiation of the test drug used, one may question this conclusion. A single well-designed clinical study demonstrates its antihistaminic, atropine-like, local anesthetic and soporific effects, but is unable to show changes in the duration of labor. On the other hand, in vitro, the drug has uterotonic activity.

Gaseous Anesthetics

Nitrous oxide, in both clinical and experimental usage, has no demonstrable effect on myometrial contractility. In vitro experiments using isolated human uterine muscle, however, show depressant effects. Similarly, ethylene and trichlorethylene exhibit practically no influence on the course of labor. Prolonged use of trichlorethylene seems to reduce the uterine contractile pattern.

The inhibitory effect of both ethyl and divinyl ether is well-documented experimentally and clinically. They diminish tonus, amplitude, and frequency of spontaneous and oxytocin-induced uterine contractions. Contractions are abolished in the lower first or upper second plane of anesthesia, and this refractory state is apparently uninfluenced by oxytocic stimulation.

Definite inhibitory effects are also documented for chloroform in both clinical and objective experiments. This agent depresses uterine contractility even in small doses and in the first plane of anesthesia, producing significant reduction in amplitude and frequency of contractions and later complete atony.

Cyclopropane depresses isolated uterine muscle strips but has no effect on the intact uterus in light anesthetic (first plane) levels. Deep cyclopropane anesthesia, on the other hand, results in progressive decrease in the frequency of uterine contractions without significantly affecting tonus or amplitude.

Halothane inhibits uterine contractions markedly and rapidly even with minimal levels. Complete obliteration of contractile response occurs with this agent. Because of its remarkable effectiveness, it is recommended as an agent of choice for tetanic uterine contractions, relaxation occurring promptly and completely. Unfortunately, severe postpartum atony which cannot be controlled by uterotonic agents may result.

Blocking Agents

Direct local inhibition of contractile patterns occurs with many of the regional blocking anesthetic agents in current use, including procaine, tetracaine, and lidocaine. Despite this, the small doses usually used clinically do not produce significant effects. Nevertheless, variations are seen with the several conduction block techniques employed in obstetrics.

With regard to the effect of spinal anesthesia on labor, discrepancies exist in otherwise apparently reliable objective studies. No consistent effect occurs with spinal anesthetic levels to sixth cervical dermatome, yet disruption of rate, rhythm, strength, and gradient of contractility are reported with lower anesthetic levels. External hysterographic study even shows significant,

prenatal life

but not persistent, rise in basal tonus, together with diminished frequency and amplitude of contractions. Further confusion is added by the correcting effect that spinal anesthesia is said to have in some cases of incoordinate uterine dysfunction. It is likely that factors other than the anesthetic agent and technique are acting here. For example, the type of labor pattern and its temporal phase are important. Spinal anesthesia before the onset of the active phase of dilatation in the first stage of labor significantly impedes progress and forestalls the expected normal progressive changes of labor. After the latent phase has ended, it should not materially influence the course of labor.

Caudal and epidural anesthesia have a negligible effect on labor unless they are inappropriately applied. Clinical studies agree that well established labor is uninfluenced. This is verified objectively by intra-amniotic pressure, external tokographic, and impedence plethysmographic studies. Anesthetic levels above thoracic-10, however, disrupt the contractility pattern and impair progress. Proper administration requires that it be given after the active phase has been entered; if earlier, labor is impeded. Many reports detail inefficient flexion and rotation of the fetal presenting part, referring to diminished voluntary expulsive force which results from abolition of the perineal reflex, rather than to any effect on myometrical function.

Paracervical block has been variously lauded for speeding cervical dilatation and decried for inhibiting uterine contractility. The former appears to be due to the rapidly advancing changes that one should anticipate as labor progresses; the latter perhaps refers to the inhibitory action of epinephrine often administered simultaneously. Tokographic study shows no consistent effect.

Pudendal block anesthesia does not interfere with uterine action. Nonetheless, combined paracervical and pudendal nerve block are alleged to speed cervical dilatation in uncontrolled clinical observations. Paravertebral lumbar sympathetic block anesthesia is described as promoting uterine contractility by increasing resting tonus and amplitude of contractions. Similarly, intra-amniotic pressure and external tokodynamometric studies show that normal contractions are improved and abnormal patterns corrected.

Muscle relaxant drugs, such as *d*-tubocurarine, gallamine, decamethonium, and suxamethonium, do not have demonstrable uterine action. Reports indicate that labor may be shortened or that the frequency and strength of *factors affecting labor*

contractions may sometimes be depressed by profound curarization. Despite this, it is generally held that cervical dilatation and uterine contractility are unaffected. Similarly, responsiveness to oxytocin is unaltered. On the other hand, diminished uterine contractile intensity is seen with succinylcholine in both subjective and objective studies.

Antispasmodics

A variably suppressing effect on uterine contractility is seen with the antispasmodic smooth muscle depressants and myovascular relaxant agents. Clinical studies of isoxsuprine show effects varying from acceleration to inhibition. In premature labor inconsistent suppression of activity occurs when the drug is used to halt labor. In vitro studies demonstrate depressant action somewhat greater than papavarine. Myometrial function is inhibited in all lower animals and in humans at term. Studies on isolated human myometrium indicate that other antispasmodic agents similarly depress the gravid uterine muscle.

Hormones

The uterus is stimulated by compounds with estrogenic activity. The intrinsic underlying capacity of the gravid uterus to contract requires the prior action of estrogenic substances. Animal and human experiments show increases in spontaneous uterine activity with stronger and more frequent contractions. However, contractions produced by estradiol are not the same as those produced by oxytocin; they are shorter, less intense, more frequent, uncoordinated, and painless. Estrogens also reduce the threshold of myometrial responsiveness to oxytocin, although this may be variable.

The role of the estrogens in uterine physiology is still incompletely clarified. It is postulated that estrogens control actomyosin synthesis and assist in the production of high-energy phosphate (ATP) to provide the wherewithal for actomyosin contraction. At the same time estrogens control both the synthesis and the activation of acetylcholine, as well as the myometrial response to it. Acetylcholine, moreover, stimulates oxytocin release, and myometrial response to oxytocin in turn can only take place if acetylcholine is present.

prenatal life

The myometrial response to progesterone is at best poorly understood. Animal experimentation indicates that progesterone inhibits uterine contractility, diminishes spontaneous contractions, and suppresses responsiveness to uterotonic agents. Human evidence is contradictory, however, gestogens appearing capable of inhibiting spontaneous contractility inconsistently. Discrepancies result from difficulties inherent in the study of myometrial tissue and from variations in techniques and in instrumentation.

Gonadotrophins inactivate rabbit myometrium, but the evidence is indirect and inconclusive. Similarly, aldosterone has an anti-progesterone effect accentuating spontaneous uterine activity.

Relaxin, a nonhomogenous water soluble ovarian extract, is alleged variously to slow or speed labor. In vitro experiments utilizing isolated human uterine muscle show no effect, however. Controlled clinical studies and objective physiologic investigations are unable to elicit consistent uterine response.

Autonomic Agents

Atropine and scopolomine are among the postganglionic cholinergic blocking agents most commonly used in obstetrics. They exert no significant influence on clinical labor in otherwise normal situations. Nevertheless, both have been shown to relax the lower uterine segment somewhat and to diminish basal tonus as well as frequency of contractions. Regulation of uterine activity in incoordinate states has also been observed.

The most widely used adrenergic blocking agents in obstetrics are the ergot alkaloids and their derivatives. In obstetrical practice today the use of these powerful uterotonic agents is limited to the postpartum period because of the excessively intensive contractions they produce in labor. The amine alkaloids, including ergonovine and the semisynthetic methylergonovine, have rapid onset of uterotonic action but minimal adrenergic activity and are effective orally as well. The dihydrogenated alkaloids, on the other hand, are essentially inactive insofar as uterine contractility is concerned, but have greater adrenergic effect than the naturally occurring amino acid alkaloids from which they are derived. Dihydroergotamine has inconsistent effect.

Large doses of the more potent adrenergic blocking agents, dibenamine and dibenzyline, have no direct effect on the uterine muscle. The imidazolines, tolazoline (Priscoline), and phentolamine (Regitine), on the other hand, produce direct uterine stimulation in many species of lower animals. The benzodioxans, piperoxan and prosympal, also stimulate the uterus directly. Although these agents should be expected to act the same way in humans, documentation of uterotonic activity has not yet been reported.

Nicotine, the most widely used ganglionic blocking agent, stimulates uterine activity in vivo but not in vitro. Whether this is due to direct stimulation or to the release of oxytocin from the neurohypophysis, a response to nicotine which is seen in the rat, has not yet been clarified in man.

Although largely replaced by hexamethonium as a therapeutic autonomic ganglionic blocking agent, the tetraethylammonium ion is mildly uterotonic, improving contractility in some forms of incoordinate labor.

Among the parasympathomimetic agents in current use, the choline ester, acetylcholine, although of limited clinical interest because of its pharmacologic instability, produces good uterine response in the intact gravid when given in large doses. In smaller doses this effect is absent. The synthetic choline derivative, carbachol, has the same effect as acetylcholine. The cholinesterase inhibitors, which prevent the rapid and continuous destruction of acetylcholine, should be expected to yield an effect upon the uterus similar to that of acetylcholine, but the uterus is apparently unaffected by physostigmine, except in very high concentrations. This is thought possibly to be due to the lack of continuous release of acetylcholine in the uterus; thus, the myometrium will not be stimulated by a drug that acts by the inhibition of cholinesterase. Similarly, neostigmine does not stimulate the pregnant uterus, except at term, despite its being erroneously labeled as an abortifacient (due to its ability to correct delayed anovulatory menstruation, an action once deemed useful in the diagnosis of early pregnancy). Among the cholinergic alkaloids, pilocarpine produces contractions in a uterus primed by estrogens.

Paradoxical effects are seen with the sympathomimetic agents. Epinephrine inhibits contractility of the pregnant uterus, and norepinephrine stimulates it. Early work using impure commercial admixtures of adrenalin, containing roughly 80 percent epinephrine and 20 percent norepinephrine, showed inhibition with low doses and stimulation with high doses. With purification of the active principles and with more objective evaluation, activity is clearly defined. Epinephrine administration diminishes frequency and

prenatal life

amplitude of contractions, while the basal tonus remains constant; rebound of uterine activity follows variably. Norepinephrine, on the other hand, initiates spontaneous contractions in vitro and augments uterine activity in vivo by increasing the frequency and amplitude of contractions; but the contractions produced often develop incoordinate activity and differ materially from those produced by oxytocin.

It has been felt that the paradoxical effects of these drugs on the uterus might be due in some cases to an action exerted directly upon the uterine muscle and in other cases to an action that may indirectly affect the uterus by stimulating the cholinergic nerves. Consequently, a sympathomimetic effect upon the uterus is obtained indirectly. With our current limited knowledge on the interrelationships of neural and hormonal function and the control of myometrial activity, this hypothesis cannot now be verified.

Other Agents

Magnesium sulphate has a definite inhibitory effect on spontaneous contractility, diminishing intensity, frequency, and tonus. The depression of uterine activity is proportional to the magnesium ion level in the blood. Clinically, the duration of labor is prolonged.

Histamine stimulates uterine muscle by direct action, but lacks uterotonic effect in clinical usage during pregnancy. The antihistamines are also mildly uterotonic. A lack of clearcut effect from dimenhydrinate has already been mentioned.

Intravenous ethyl alcohol analgesia slows or stops normal spontaneous labor, and even relaxes the tetanically contracted uterus, but does not affect oxytocin-induced labor. Alcohol may actually inhibit the release of oxytocin from the pituitary gland.

Digoxin causes an in vitro increase in frequency of contraction with failure of complete relaxation between contractions, as does ouabain and strophanthin G in vivo. Clinical labor, however, is unaffected. It has been conjectured that the uterotonic effect of the cardiac glycosides may be functionally related to similar pharmacologic activity of sparteine sulphate.

Bradykinin is inconsistently uterotonic in the rat but not in the human being. Nevertheless, this drug effectively blocks spontaneous and oxytocin-induced contractions of human muscle strips, decreasing the frequency and amplitude. The degree of relaxation induced by bradykinin is proportional to the dose. The contradictory reports on this and other drugs indicate the obvious need for additional study.

Conclusion

I have raised more questions than I have answered here. The prevailing state of ignorance with regard to basic uterine physiology has been all too clearly illustrated. Where controversies exist, I have tried briefly to point them out. These must await resolution by refined, objective techniques of study under ideal experimental conditions. As critical investigative techniques for the study of complex labor phenomena are developed, correlation of data derived from meaningful experiments will hopefully clarify the physiology of the pregnant uterus and the pharmacologic effects of exogenously administered agents.

chapter XI

The Effect of Intrauterine Hypoxia/Asphyxia on the Surviving Child

Kenneth R. Niswander, M.D.

Traditionally, an obstetrician has considered his work a success if the mother at the conclusion of the delivery is healthy and if the fetus is born alive and survives the neonatal period. A living infant has been the desired and only measurable endpoint of obstetrical care and delivery. A low perinatal mortality rate, therefore, was equated with successful management of a particular obstetrical problem, and a high perinatal death rate with a failure of therapy.

Although Little[1] in 1862 stressed the association of "abnormal parturition, difficult labor, premature birth, and asphyxia neonatorum" with cerebral palsy, thus precipitating a long-standing debate in the literature concerning the etiological factors involved, it was in the 1950's when this generation of obstetricians had its attention focused on the possible relationship between obstetrical factors and neurologic damage. In 1951, Lilienfeld

---

This study was supported by USPHS Grant NB-02404, NIH, and by similar grants to the other participating institutions. The collaborative study of cerebral palsy, mental retardation, and other neurological and sensory disorders of infancy and childhood is supported by the National Institute of Neurological Diseases and Blindness. The following institutions participate: Boston Lying-in Hospital; Brown University; Charity Hospital, New Orleans; Children's Hospital of Buffalo; Children's Hospital of Philadelphia; Children's Medical Center, Boston; Columbia University; Johns Hopkins University; University of Minnesota; Medical College of Virginia; New York Medical College; Pennsylvania Hospital University of Oregon; University of Tennessee; Yale University; and the Perinatal Research Branch, National Institute of Neurological Diseases and Blindness.

prenatal life

and Parkhurst,[2] in a classic paper on the etiology of cerebral palsy, postulated that intrauterine hypoxia or asphyxia might cause varying fetal outcomes depending upon the severity of the insult. Severe intrauterine embarrassment of the fetus would cause fetal death, while a less severe insult might result in a live fetus which did not survive the neonatal period. A yet smaller damage. If a fetus avoided major neurologic damage, he might still show evibut significant insult might result in cerebral palsy or other severe neurologic dence of mild brain damage. These authors called this range of fetal outcomes a "continuum of reproductive wastage." No longer could the obstetrician be satisfied with a live baby. It became necessary for him to consider the effect of his treatment not only on the fetus but also on the child as he developed. Thus was born the need for a longitudinal approach to obstetrical research. Long-term observation of the infant became a necessity.

That intrauterine as well as extrauterine hypoxia or asphyxia can cause neurologic damage is unquestioned. In classical animal experiments using asphyxiated guinea pig fetuses Windle[3] clearly demonstrated severe neurologic dysfunction in a high proportion of surviving pigs. It is important to note, however, that "neurologic symptoms as a rule failed to persist in marked form throughout life and were often only transient." Muller and Graham[4] reported that six of eight children who survived accidental intrauterine carbon monoxide poisoning were spastics. The authors did not state, however, how many of these children were of low birth weight, a factor known to be a major determinant of neurologic outcome. Rosen,[5] in a beautiful experimental preparation in the guinea pig, showed that with maternal asphyxia the fetal EEG pattern quickly flattened and returned to a normal configuration appreciably later than the fetal heart rate returned to normal. Rosen did not investigate ultimate neurologic condition of the fetuses.

There is no doubt then that severe intrauterine asphyxia can be associated with brain damage, which may be transient or may extend into later life. Certain questions, however, remain unanswered. Is the asphyxia-brain damage relationship clinically a frequent one? Is the neurologic effect of asphyxia clinically transient, or is it likely to persist? Is intrauterine asphyxia the major cause of a significant percentage of the neurologically abnormal population—the cerebral palsied individuals, the mentally retarded, the minimal brain damage-behavioral problems? We chose an epidemiologic approach to attempt to answer these questions.

intrauterine hypoxia/asphyxia

Material and Method

The Collaborative Project for the Study of Cerebral Palsy is a prospective study of large numbers of gravida. In addition to detailed prenatal, labor, and delivery data, an evaluation of the child at birth is provided by Apgar scores and during the nursery course by frequent examinations. At four and twelve months of age neurological evaluation is done. At eight months the psychologic examination provides an estimate of motor function and to some degree of mental development. At four years of age a Stanford-Binet I.Q. score is determined, together with estimates of fine motor, gross motor, concept formation, and behavioral function. Later examinations beyond four years of age are also planned. The protocol has been described elsewhere.[6]

Records of 40,262 cases obtained from thirteen institutions were available for detailed study, adequate time having elapsed from delivery to permit at least a one-year followup of surviving children for neurologic evaluation. Only pregnancies resulting in a single fetus of 500 grams or more were included. To study the effect on the fetus and growing child of hypoxia/asphyxia, project nurses abstracted data from the charts of patients with the following diagnoses: abruptio placentae (566 patients), placenta previa (232 patients), and prolapse of the umbilical cord, both frank (166 patients) and occult (166 patients). In the absence of an exact diagnosis of fetal hypoxia, which cannot be made with accuracy at the present time, we chose this group of patients in whom the risk of hypoxia seemed substantial. For certain tabulations we analyzed separately the study patients who suffered severe shock or in whom substantial slowing of fetal heart rate was noted. In this way we hoped to define more clearly a group of patients in whom fetal hypoxia was even more likely to have been present.

Results

Table 1 totals the perinatal deaths among our study patients. Clearly, these diseases increased the risk of perinatal death above the usual incidence.

Table 2 lists the percentages of babies with depressed one-minute Apgar scores of 0 to 6. The control babies are the total Collaborative Project population. The study babies, both low birth weight and mature weight, had

prenatal life

an increased incidence of low one-minute Apgar score, a highly significant difference ($\chi^2 = 230.55$, $df = 1$, $p < .01$).

Table 3 shows similar relationships for the five-minute Apgar score. The differences between the study and the control babies again are highly significant, both for the low birth weight and the mature birth weight groups ($\chi^2 = 109.21$, $df = 1$, $p < .01$). Do these findings permit us to say that the babies as a group were hypoxic or asphyxiated in utero?

Table 1. Perinatal Deaths

	Percentage	
	Low Birth Weight (<2500 grams)	Mature Birth Weight (2500+ grams)
placenta previa	34.7 (34/98)	7.4 (10/133)
abruptio placentae	47.0 (119/253)	10.6 (33/312)
cord prolapse—frank	55.2 (32/58)	14.2 (15/106)
cord prolapse—occult	36.8 (7/19)	4.2 (6/146)
total hypoxia	44.9 (192/428)	9.2 (64/697)
control population	17.7 (311/1,762)	1.1 (166/14,635)

Table 2. Depressed One-Minute Apgar Score

	Percentage	
	Low Birth Weight (<2500 grams)	Mature Birth Weight (2500+ grams)
placenta previa	64.5 (58/90)	36.4 (44/121)
abruptio placentae	66.0 (103/156)	34.3 (91/265)
cord prolapse—frank	82.4 (28/34)	60.9 (53/87)
cord prolapse—occult	64.3 (9/14)	35.5 (48/135)
all hypoxia	67.3 (198/294)	38.8 (236/608)
control population	34.3 (486/1,417)	19.1 (2,536/13,257)

Table 3. Depressed Five-Minute Apgar Score

	Percentage	
	Low Birth Weight (<2500 grams)	Mature Birth Weight (2500+ grams)
placenta previa	46.2 (42/91)	8.6 (11/127)
abruptio placentae	42.7 (67/157)	8.2 (21/257)
cord prolapse—frank	51.5 (17/33)	19.8 (18/91)
cord prolapse—occult	23.1 (3/13)	8.1 (11/135)
all hypoxia	43.9 (129/294)	10.0 (61/610)
control population	16.9 (244/1,442)	3.9 (525/13,553)

The measuring of the effect of intrauterine hypoxia or asphyxia is complicated by the difficulty of making the actual diagnosis of hypoxia. pH determination of the fetal blood, base deficit, O_2 and CO_2 tensions, and other blood chemistry determinations may reflect presumed asphyxia. Likewise, certain fetal heart rate patterns are thought to be due to hypoxia.[7] The exact duration of the insult and the severity of the insult, however, are usually unknown. It has been our feeling that the Apgar scores perhaps very accurately summate the effects of intrauterine factors on the fetus, at least those which occurred just prior to delivery. Depressed Apgar scores thus led us to conclude that our babies as a group had probably been exposed to an unfavorable intrauterine environment and were asphyxiated at birth.

Figure 1 is a display of the results of the four-month examination on which the babies are scored as neurologically normal, neurologically suspect, or neurologically abnormal. 1.9 percent of the low birth weight study babies were considered abnormal, while 2.3 percent of the total Collaborative Project population babies were scored similarly. Among the mature babies 1.2 percent of the study group and 2.2 percent of the control population were abnormal. Thus, no significant difference could be detected between the two groups of babies at four months of age in spite of the marked differences at birth.

At the time of the eight-month examination (Fig. 2), 15.1 percent of the study low birth weight babies had the lowest category of mental score, while 8.5 percent of the total Collaborative Project population were similarly scored ($\chi^2 = 13.55$, $df = 3$, not significant). Among the mature babies, the figures were 2.1 percent in the study babies and 1.3 percent in the controls ($\chi^2 = 20.27$, $df = 3$, $.01 < p < .05$). Findings similar to these mental score results are noted on the eight-month motor scores (Fig. 3); 18.9 percent of the low birth weight study babies were scored in the lowest grouping, compared to 13.4 percent of the control prematures ($\chi^2 = 8.42$, $df = 3$, not significant), and 2.8 percent versus 2.2 percent among the mature infants ($\chi^2 = 2.35$, $df = 3$, not significant). The examiner's clinical impression of the overall function of the infant at eight months is scored as global impression (Fig. 4). 12.8 percent of the low birth weight study babies were considered abnormal, and 11.7 percent of the control babies were abnormal. Among the mature birth weight children, 2.1 percent and 2.2 percent were considered abnormal, respectively, obviously insignificant differences.

prenatal life

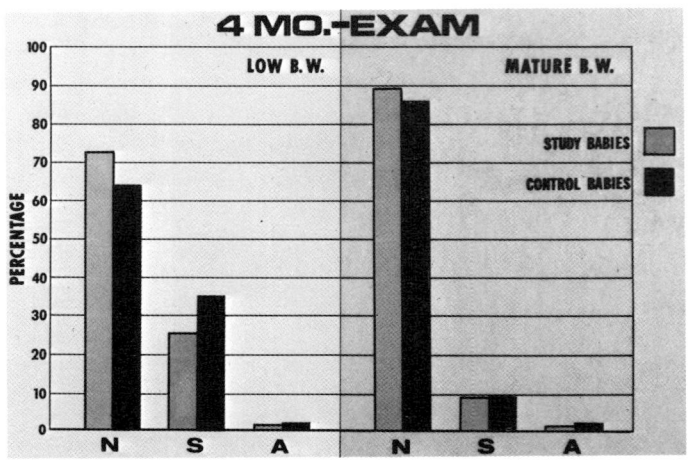

Figure XI-1. N = neurologically normal, S = suspect, A = abnormal.

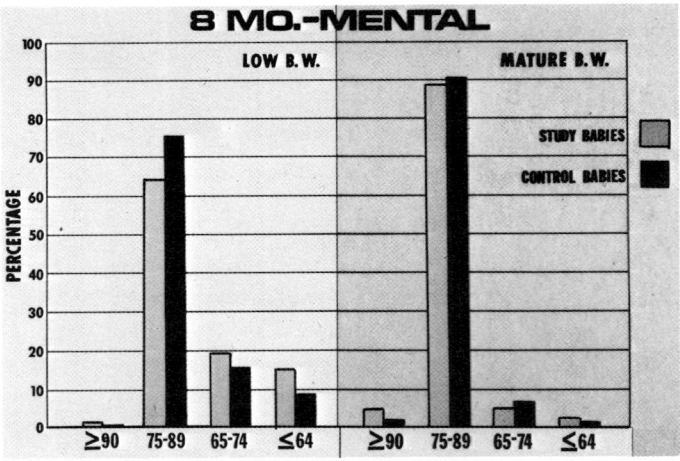

Figure XI-2. ≥ 90 = highest mental score, ≤ 64 = lowest mental score.

intrauterine hypoxia/asphyxia

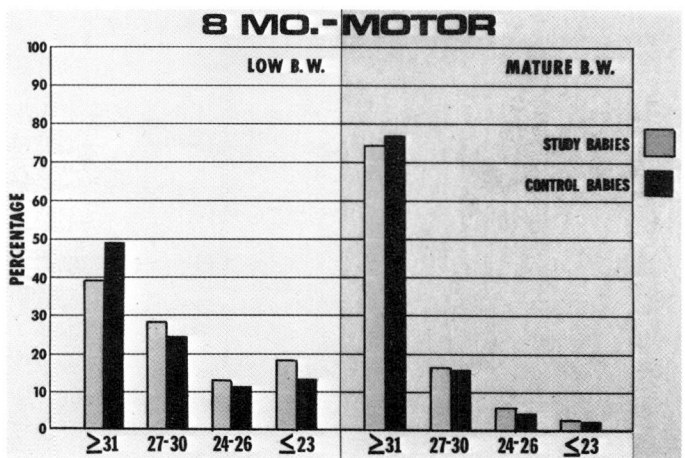

Figure XI-3. ≥ 31 = highest motor score, ≤ 23 = lowest motor score.

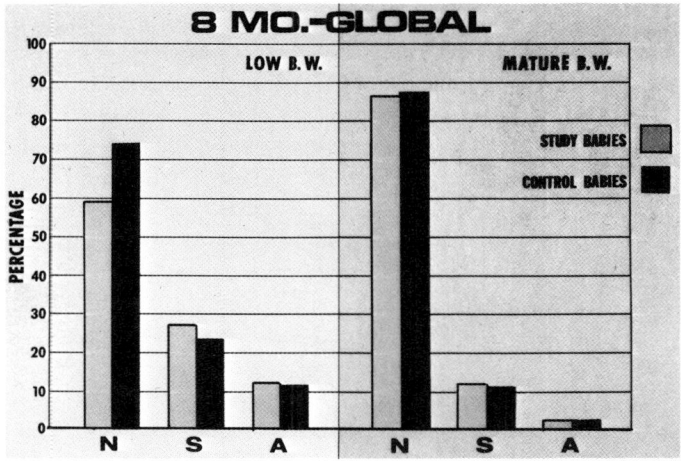

Figure XI-4. N = neurologically normal, S = suspect, A = abnormal.

prenatal life

Figure 5 shows the results of the one-year conventional neurologic examination. Among the low birth weight study infants 4.8 percent were frankly abnormal and 78.3 percent were normal. Among all the Collaborative Project low birth weight infants 4.9 percent were abnormal and 84.0 percent were normal. In the mature group of infants 1.3 percent of the study babies and 1.5 percent of the control babies were abnormal, while the percentages of normal children were 93.3 percent and 92.0 percent, respectively. No statistical significance in the differences was found.

We can conclude that the study babies were depressed at birth probably as a result of hypoxia/asphyxia. Neurologic evaluation to the age of one year failed to demonstrate any differences between the study and control babies. At all examinations the low birth weight group of babies fared notably poorer than the mature birth weight group whether they were hypoxic at birth or not, and prematurity would appear to be a major factor in determining risk of neurologic deficit. To date only 263 of the surviving children have been examined at four years, a number insufficient to allow dependable conclusions regarding mental status at this age.

In an attempt to delineate more precisely the relationship between severe asphyxia or hypoxia and neurologic development, we chose from the placenta previa and the abruptio placentae populations two sets of patients: the first was composed of the mothers who were recognized as suffering from severe shock, and the second comprised those with no shock. There were seventy-eight gravida in the shock group and 613 in the no-shock category. By comparing the outcomes in the infants born to these mothers, we hoped to be able to differentiate the effects of severe hypoxia/asphyxia against mild or no asphyxia. Figure 6 shows a significant (low birth weight: $\chi^2 = 9.4, df = 1, p < .01$; mature birth weight: $\chi^2 = 40.53, df = 1, p < .01$); and expected relationship between severity of shock and perinatal death. Among the low birth weight group, 38.1 percent of the shock infants survived, while 63.1 percent of the no-shock babies survived. Similar figures among the mature babies were 61.1 percent and 93.6 percent. Severe maternal shock also correlated well with depressed one- and five-minute Apgar scores. At one-minute (Fig. 7) 82.8 percent of the low birth weight study infants had a depressed Apgar score, while 62.1 percent of the no-shock infants had a similarly depressed score ($\chi^2 = 4.72, df = 1, p < .05$). Among the mature birth weight groupings the percentages were 56.5 and 33.2, re-

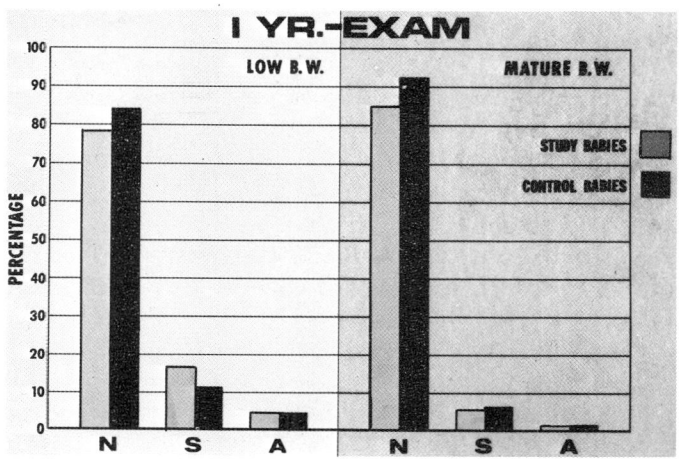

Figure XI-5. N = neurologically normal, S = suspect, A = abnormal.

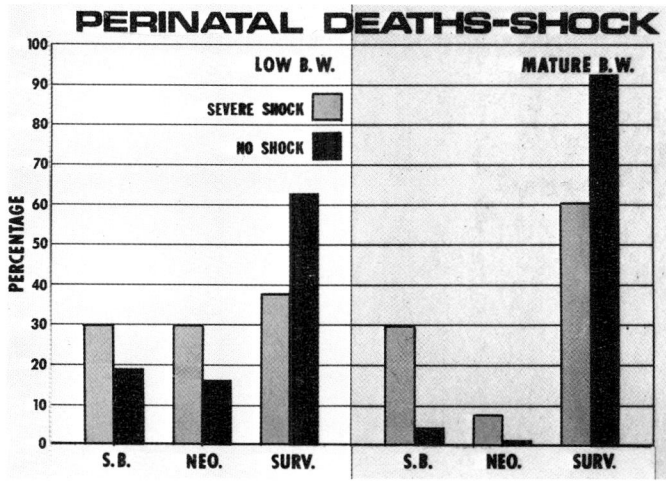

Figure XI-6. S.B. = fetal deaths, Neo. = neonatal deaths, Surv. = survived neonatal period.

prenatal life

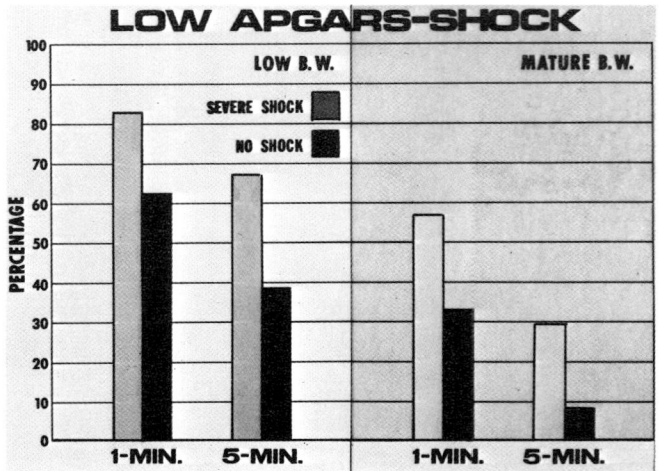

Figure XI-7. Percentage Apgar 0 to 6.

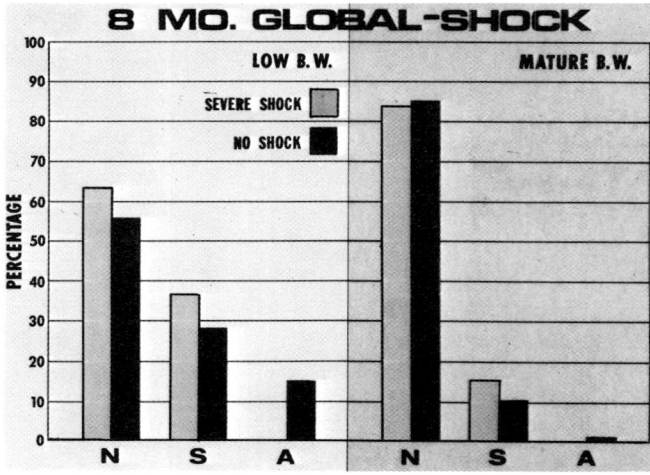

Figure XI-8. N = neurologically normal, S = suspect, A = abnormal.

intrauterine hypoxia/asphyxia

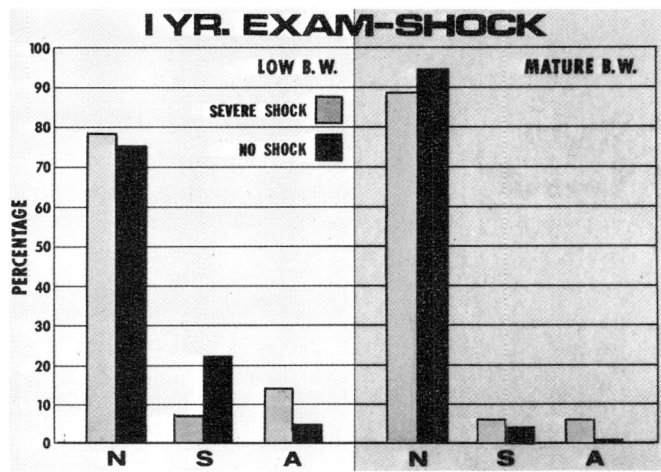

Figure XI-9. N = neurologically normal, S = suspect, A = abnormal.

spectively ($\chi^2 = 5.12$, $df = 1$, $p < .05$). At five-minutes the differences were even more obvious with 66.7 percent of the low birth weight study infants depressed, and 38.6 percent of the control babies depressed ($\chi^2 = 8.33$, $df = 1$, $p < .01$). Of the mature group babies, 29.2 percent of the shock infants were depressed, and only 8.8 percent of the no-shock infants were depressed ($\chi^2 = 10.12$, $df = 1$, $p < .01$). No significant differences were noted in the global score at eight months (Fig. 8). At one year of age (Fig. 9), 14.3 percent of the premie shock babies were abnormal, while 4.8 percent of the no-shock low birth weight infants were abnormal ($\chi^2 = 2.18$, $df = 1$, not significant). In the mature babies 5.9 percent of the study babies and 1.1 percent of the control babies were abnormal ($\chi^2 = 2.88$, $df = 1$, not significant). The inverse relationships on these two examinations is apparently due to small numbers of cases, and the differences between the groups are not significant.

The fetal heart rate was sampled in the Collaborative Project patients at very frequent intervals. The protocol directed the nurse observer to count the fetal heart rate immediately after the cessation of a contraction. When we abstracted the charts for this investigation, using only the group diag-

prenatal life

nosed as having prolapse of the umbilical cord, we categorized the fetus as having had a heart rate counted at less than 100 on one or more occasions or having had no heart rate recorded below 100. Figures 10 and 11 show the relationship between such a division of fetal heart rates and outcomes. Figure 10 shows no correlation between a low fetal heart rate and one-minute Apgar scores, either in the low birth weight group or the mature birth weight group. 67.1 percent of the prematures with a low fetal heart rate were depressed at one minute, while 67.7 percent with no fetal heart rate drop were depressed. The corresponding figures among the mature babies were 43.1 percent and 36.3 percent. At five minutes the percentages of depressed Apgars were 46.1 percent and 44.1 percent in the prematures and 12.9 percent and 10.6 percent in the mature babies. At eight months of age 11.9 percent of the low birth weight babies with a drop in fetal heart rate were abnormal, while 14.4 percent of those without a drop in fetal heart rate were abnormal. Comparable figures in the mature weight group were 1.0 and 2.7 percent. At one year of age (Fig. 11) 5.6 percent of the study prematures and 3.8 percent of the control prematures were abnormal. In the mature group 1.5 percent of the study group and 1.2 percent of the control group were abnormal. Obviously, frequent sampling of the fetal heart did not allow a correlation between depressed fetal heart rate and Apgar scores or neurologic outcome to the age of one year. It should be noted that continuous fetal heart rate monitoring as reported by Hon[7] and Caldeyro-Barcia *et al.*[8] was not in the protocol of the Collaborative Project. Continuous monitoring would undoubtedly have identified depressed infants more accurately.

Discussion

We posed certain questions earlier that we should now try to answer. It seems certain from the work of others that severe intrauterine asphyxia can be associated with brain damage, but we asked ourselves: Is the asphyxia-brain damage relationship clinically a frequent one? Certainly from the data we have presented, one can tentatively conclude that motor dysfunction following asphyxia is not common. Other data we have collected from the Collaborative Project on potentially asphyxiating diseases support the observed lack of effect of asphyxia on motor function. Maternal organic heart

intrauterine hypoxia/asphyxia

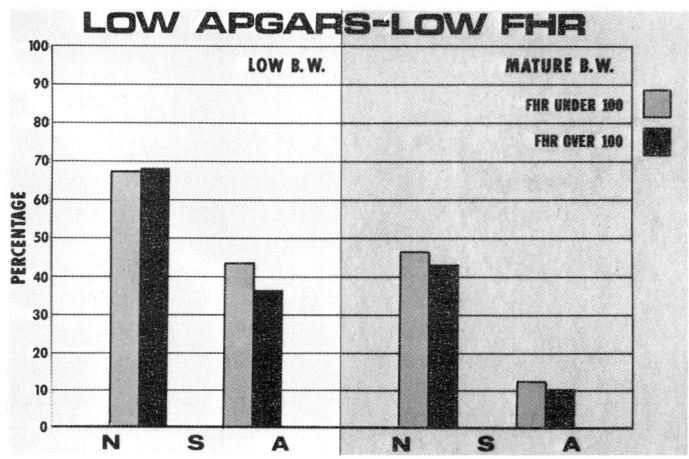

Figure XI-10. Percentage Apgar 0 to 6. Low FHR = FHR recorded one or more times less than 100 beats per minute.

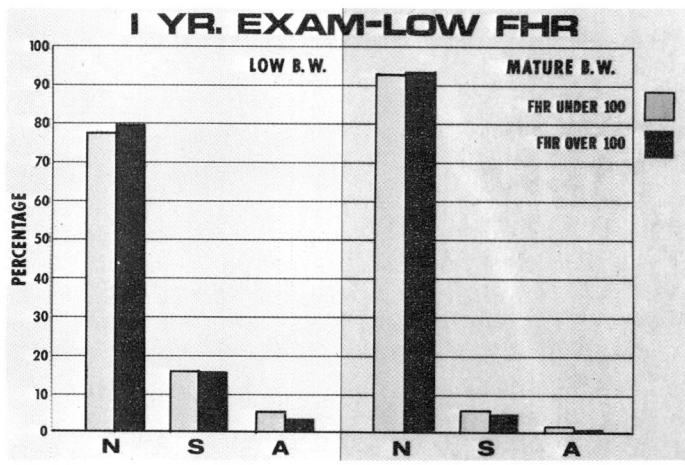

Figure XI-11. N = neurologically normal, S = suspect, A = abnormal.

prenatal life

disease in pregnancy was associated with a moderate increase in perinatal death risk and moderate depression of the Apgar scores of the babies delivered by women with this pathology.[9] No evidence of increased risk of neurologic damage at eight months or one year of age was noted, however. A separate analysis of the gravida who suffered heart failure during pregnancy and those who did not have cardiac decompensation did not change the results. Maternal asthma was related to depressed Apgar scores in the infants at birth, but the percentage of abnormal children on the eight-month and one-year exams were almost identical to the total Collaborative Project population.[10] Analysis of the patients with a diagnosis of rupture of the marginal sinus revealed similar results in the infants—depressed Apgar scores but no increase in the number of abnormal babies up to one year of age.[11]

Additional evidence to support our thesis may be adduced from a study of elective induction of labor, a procedure thought by some potentially to subject the fetus to hypoxia. D'Esopo et al.[12] and Wells[13] showed that elective induction of labor increased the number of infants with depressed Apgar scores. And Niswander et al.[14] showed, in a controlled study of 131 babies born after labors electively induced and 147 babies born after labors which began spontaneously, that respiratory depression during the neonatal period was nearly twice as frequent among the study babies. At four years of age, however, we[14] could demonstrate no significant neurologic differences between the two groups.

A second question we asked ourselves was: Is the neurologic effect of asphyxia transient or is it likely to persist? It was not possible to tabulate nursery exams in the present study, but in an earlier study with smaller groups of babies we showed not only that the Apgar scores were depressed following placental abruption, placenta previa, and cord prolapse, but also that neurologic abnormalities diagnosed during the nursery period were three to four times as frequent in the study babies as in the total Collaborative Project population. Yet on later examinations these differences disappeared.[15] We speculated that "recovery" of brain function might have occurred. Windle[3] suggested such an occurrence in guinea pigs. MacKinney[16] showed a slight decrease in the mean intelligence quotient of children born asphyxiated, as compared with a control population, but this difference seemed gradually to have disappeared by age five or six years. It seems likely that brain damage can be transient, either because the pathology causing the damage (edema?) resolved, or because new neuronal pathways permitted bypassing of abnormal areas.

A third question we posed was: Is intrauterine asphyxia a major cause of a significant percentage of the neurologically abnormal population? Our findings at the present time do not permit an answer to that question. Others have answered it, however, and we might analyze these studies briefly. Lilienfeld and Parkhurst implicated the following obstetrical conditions in order of frequency as etiologic agents of cerebral palsy: placenta previa, abruptio placentae, prolapse of the cord, and maternal dystocia. Prematurity, which is strongly related to subsequent neurologic dysfunction, occurs with decreasing frequency in these four conditions—most often with previa and abruption, less with cord prolapse, and least often with dystocia. It seems possible that Lilienfeld and Parkhurst were measuring primarily an effect of prematurity rather than asphyxia.

Eastman et al.[17] conducted a similar study on the etiology of cerebral palsy. Although these authors found a relationship between intrauterine hypoxia and palsy, the correlation was much less striking than Lilienfeld and Parkhurst had suggested. Bacola et al.[18, 19] studied the effect of asphyxia on premature infants and concluded that intrauterine or perinatal asphyxia has little effect on subsequent brain development. Postnatal asphyxia in the nursery, however, was found to be much less benign in this regard. That severe shock in our group of mothers apparently exerted so little effect on neurologic development strengthens the conclusion that intrauterine asphyxia is not a very common cause of major neurologic damage.

This is an ongoing study, and one is hesitant to draw firm conclusions that later may require revision. We might tentatively, however, suggest the following from the data we have presented: (1) we have confirmed the obvious—that intrauterine hypoxia/asphyxia is related strongly to the perinatal death rate and has an adverse effect on the immediate condition of the surviving newborn; (2) we have been unable to substantiate the opinion that intrauterine hypoxia is a common cause of cerebral dysfunction; and (3) we can confirm from our data the frequency with which low birth weight is an antecedent of brain damage. Research toward the eradication of premature labor is perhaps the single most important unsolved problem in obstetrics today.

prenatal life

References

1. Little, W. J.: On the influence of abnormal parturition, difficult labours, premature birth, and asphyxia neonatorum, on the mental and physical condition of the child, especially in relation to deformities. *Trans Obstet Soc, London, 3*: 293, 1862.
2. Lilienfeld, A. M., and Parkhurst, E.: A study of the association of factors of pregnancy and parturition with the development of cerebral palsy: A preliminary report. *Amer J Hyg, 53*: 262-283, 1951.
3. Windle, W. F.: Structural and functional alterations in the brain following neonatal asphyxia. *Psychosom Med, 6*: 155-156, 1944.
4. Muller, G. L., and Graham, S.: Intrauterine death of the fetus due to accidental carbon monoxide poisoning. *New Eng J Med, 252*: 1075-1078, 1955.
5. Rosen, M. G.: Effects of asphyxia on the fetal brain. *Obstet Gynec, 29*: 687-693, 1967.
6. Berendes, H.: *Research Methodology and Needs in Perinatal Studies*, ed. S. Chipman, A. Lilienfeld, B. Greenberg, and J. Donnelly, Springfield, Thomas, 1966, pp. 118-138.
7. Hon, E. H.: Electronic evaluation of the fetal heart rate: VI. Fetal distress—a working hypothesis. *Amer J Obstet Gynec, 83*: 333-353, 1962.
8. Caldeyro-Barcia, R., Mendez-Bauer, C., Poseiro, J. J., Escarcena, L. A., Pose, S. V., Bieniarz, J., Arnt, I., Gulin, L., and Althabe, O.: *The Heart and Circulation in the Newborn and Infant*, ed. D. E. Cassels, New York, Grune & Stratton, 1966, pp. 7-36.
9. Niswander, K. R., Berendes, H., Deutschberger, J., Lipko, N., and Westphal, M. C.: Fetal morbidity following potentially anoxigenic obstetric conditions: V. Organic heart disease. *Amer J Obstet Gynec, 98*: 871-876, 1967.
10. Gordon, M., Niswander, K. R., and Kantor, A. G.: Fetal morbidity following potentially anoxigenic obstetric conditions: VII. Maternal asthma. *Amer J Obstet Gynec*. In Press.
11. Niswander, K. R., Berendes, H. W., Deutschberger, J., Weiss, W., Lipko, N., and Kantor, A. G.: Fetal morbidity following potentially anoxigenic obstetric conditions: VI. Rupture of the marginal sinus. *Amer J Obstet Gynec, 100*: 862, 1968.
12. D'Esopo, D. A., Moore, D. B., and Lenzi, E.: Elective induction of labor. *Amer J Obstet Gynec, 89*: 561-567, 1964.
13. Wells, J.: Effect on the newborn of induced versus spontaneous labor. *Obstet Gynec, 26*: 580-584, 1965.
14. Niswander, K. R., Turoff, B. B., and Romans, J.: Developmental status of children delivered through elective induction of labor. *Obstet Gynec, 27*: 15-20, 1966.
15. Niswander, K. R., Friedman, E. A., and Berendes, H.: Do placenta previa,

abruptio placentae, and prolapsed cord cause neurologic damage to the infant who survives? *Develop Med Child Neurol* (Unpublished manuscript).
16. MacKinney, L. G.: *Neurological and Psychological Deficits of Asphyxia Neonatorum*, ed. W. F. Windle, Springfield, Thomas, 1958, pp. 195-218.
17. Eastman, N. J., Kohl, S. G., Maisel, J. E., and Kavaler, F.: The obstetrical background of 753 cases of cerebral palsy. *Obstet Gynec Survey*, 17: 459-500, 1962.
18. Bacola, E., Behrle, F. C., deSchweinitz, L., Miller, H. C., and Mira, M.: Perinatal and environmental factors in late neurogenic sequelae: I. Infants having birth weights under 1,500 grams. *Amer J Dis Child*, 112: 359-368, 1966.
19. ———: Perinatal and environmental factors in late neurogenic sequelae: II. Infants having birth weights from 1,500 to 2,500 grams. *Amer J Dis Child*, 112: 369-374, 1966.

chapter XII

The Selection of Indicators of Fetal Circumstance

Karlis Adamsons, M.D., Ph.D.

The relative constancy of the fraction of human infants born under suboptimal conditions, as judged by Apgar score, is indicative of the inadequacy of the standard techniques applied in the surveillance of the intrauterine patient during the course of labor and delivery. There is little disagreement that acute asphyxia of considerable degree is associated with characteristic and readily detectable changes in fetal heart rate and, in most instances, also with passage of meconium. The more gradually developing disturbances in fetal homeostasis, on the other hand, are not necessarily expressed in these phenomena until marked compromise of the intrauterine patient is reached. Since a number of clinical situations belong to this category, considerable effort has been made to seek other indicators that would allow the detection of incipient disorders.

The relative inaccessibility of the fetus and the lack of knowledge which functions of the fetal organism deserve particular consideration have meant that an entirely satisfactory method for surveillance of fetal safety is still to be developed. Ideally, the variable studied should reflect a vital function of the least tolerant system to disturbances in the composition of the intra and extracellular environments. The function chosen should be one in which the disturbance is irreversible rather than contingent upon the prevalence of the adverse factor. For example, even if glomerular filtration rate

Supported in part by USPHS Research Grants HD 118 and GM 0906.

of the fetus were strongly associated with the degree of fetal oxygenation, it might not be a suitable indicator of fetal safety, because minor forms of asphyxia of no consequence to the integrity of the fetus could lead to substantial alteration in this variable.

According to generally held views, the central nervous system of the mature fetus is most susceptible to asphyxia. There appears to be an inverse relationship between the tolerance of a given tissue to disturbances in homeostasis and its resting metabolic rate. The myocardium in this regard occupies a unique position, because one is not concerned with the ability of this tissue to survive oxygen deprivation but with its ability to maintain perfusion of other organs.

It still remains to be elucidated whether the assessment of biologic function for surveillance purposes is best achieved by physical or by chemical means. Although it can be stated a priori that the change in function must be preceded by change in the composition of intra or extracellular environments, it does not necessarily follow that the resolution power of biochemical methods compares favorably with that of methods applicable to physical variables reflecting the performance of altered tissues. An example of this would be the superiority of diagnosing fatigue of the central nervous system by a performance test designed to assess the power of association over the biochemical analysis of the blood entering or leaving the central nervous system. Although it can be proposed that there is a specific constituent that accurately reflects the performance characteristic of the system, it may either be unknown or difficult to quantitate. A further advantage of biophysical variables is the relative facility of their continuous monitoring. Examples of this include fetal heart rate, electrocardiogram, and fetal temperature.

Among the limitations of biophysical indicators of fetal circumstance perhaps the most serious is the fact that a normal signal may be observed in the presence of an irreversibly altered state of tissues. Furthermore, the physical phenomenon may originate in an organ which yields a normal output in the presence of a number of abnormal but opposing factors. This particularly pertains to the fetal heart, which responds to direct stimuli of the central nervous system, to humorally transmitted substances, and to a reduction in the availability of energy. It has been demonstrated, at least in the fetus of the subhuman primate, that severe degrees of disturbance in homeostasis may still be associated with normal heart rate and blood pressure. Similar reservations also appear to apply to the electrical activity of the heart as detected by ECG. Recent observations suggest that the rate of change in the composition of the intra and extracellular fluid may be important in determining the changes in the output of the effector organs. It is likely that some of these limitations regarding the reliability of biophysical indicators of fetal circumstance could be overcome by resorting to evoked responses.

The principal virtue of a biochemical method in assessing the fetal condition is the relative independence of the properly selected variable from compensatory influences. As a drawback one must list the difficulty in obtaining samples and the necessity to assume that the composition of blood or other available fluids reflects changes in the intra and extracellular compartments of those organs which are most susceptible to disturbances in homeostasis. This is likely to apply to blood, since equilibration between tissues and blood is rapid. Furthermore, organs in which energy transformation rates are high, such as myocardium and the central nervous system, derive a large proportion of the total cardiac output even under conditions of asphyxia. In this regard, amniotic fluid is a far less satisfactory source, because of the low rate with which changes in fetal homeostasis alter the composition of this fluid. Difficulties may result in the interpretation of data, because information may be lacking regarding the degree of departure from normality in composition that is still compatible with normal performance of the tissue and unimpaired recovery. A similar comment pertains to situations in which the factors leading to an abnormal composition of body fluids are unknown; for instance a low pH of fetal blood may be of little consequence if it results from a high CO_2 tension in a well-oxygenated mother. Most of these limitations regarding the validity of biochemical indicators can be overcome by a better understanding of the associated variables and by increased reliance upon the difference in concentration of a constituent between the maternal and fetal compartment rather than on its absolute value.

A shortcoming which is nearly characteristic of biochemical assessment of any system is the necessity to remove the specimen for analysis, which essentially precludes a continuous monitoring of the chosen variable. Indirect methods are available to measure the tension of oxygen and carbon dioxide in fetal tissues. Unfortunately, they are of little value in the management of the human fetus. Even if technical difficulties of determining accurately the tension of gases by an electrode were removed, the limitation would still pertain that the tension of O_2 or CO_2 in the extracellular fluid of peripheral tissues may bear little relationship to the availability of energy to vital organs. Thus, it appears that examination of capillary blood of the fetus, a technique introduced by Saling, will remain in the foreseeable future the procedure of choice in the surveillance of the intrauterine patient.

Of the constituents that can be accurately quantitated in minute blood samples, the components reflecting the acid-base state are of particular value. It is not because biologic systems are uniquely susceptible to changes in acid-base state, but because the activity of the hydrogen ion is one of the best indirect measures of the availability of biologically useful energy for cell function. This is due to the fact that in the absence of oxygen, when pyruvic acid becomes the principal hydrogen acceptor from the reduced DPN, a large amount of hydrogen ion is formed by the reduction product lactic acid. Although only a small fraction, usually not more than a small percentage, appears in blood as free hydrogen ion, it represents a large change in the concentration of this constituent. Under conditions of total asphyxia, the rate of rise in hydrogen ion concentration in the mature unanesthetized fetal monkey is approximately 20 nM/Lx min. corresponding to a fall in pH of about 0.1 per minute. Since the accuracy of pH measurement is better than plus or minus 0.01, it is expected that even transient interruption in placental exchange should be detectable by accessing serially the hydrogen ion concentration of blood. It remains to be elucidated whether fetal acidosis develops during a normal uterine contraction. A certain stability of the internal environment is provided by the amount of oxygen stored in blood and by the decrease in the ionization of the imidazole groups, a principal hydrogen ion donor of the hemoglobin molecule, when it is changed from the oxy to reduced form. Beside the facility with which the concentration of hydrogen ion can be determined, this variable offers a further advantage for clinical monitoring of the fetus. This consists of the latency with which pH returns to normal values after establishment of normal oxygen supply. As can be seen from Figure 1, abnormal pH is still present as long as thirty minutes after the initial insult. Similar although less persistent changes in pH are also observed in the sheep fetus following transient occlusion of the umbilical cord. The changes in the concentration of excess hydrogen ion (identical with base deficit) during asphyxia and recovery are similar to those described for pH.

indicators of fetal circumstance

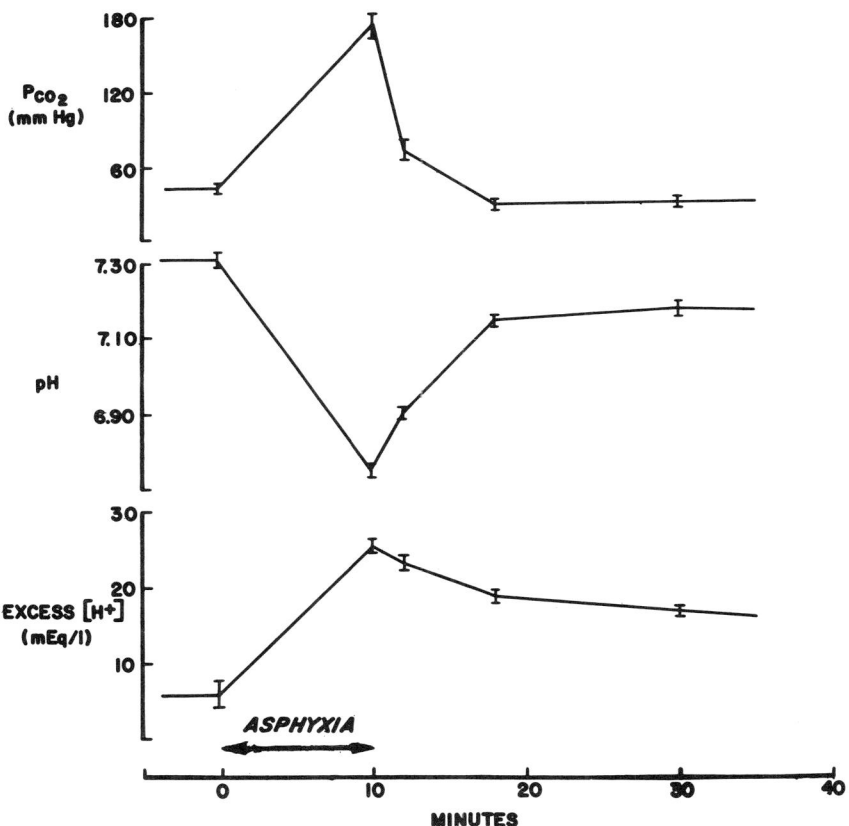

Figure XII-1. The effect of asphyxia upon PcO$_2$, pH, and Excess[H+] in newborn monkeys. (Redrawn from data, Adamsons *et al.*, *J Pediatrics*, 65: 807, 1964.)

Due to the slow clearance of hydrogen ion the elevation of levels in the recovery period is even more pronounced. The advantage of excess of hydrogen ion as indicator of oxygen availability to the fetus is somewhat minimized by the more time-consuming procedure for the determination of this variable and the lesser degree of accuracy of measurement. Values of excess acid in fetal blood may also be reflecting the metabolic acidosis in the mother, a not uncommon situation in the laboring patient. This factor has been particularly emphasized by Beard and coworkers, who have suggested that the difference in excess hydrogen ion concentration between mother and fetus, rather than the absolute value of fetal blood, is a more reliable indicator of fetal condition. There is little doubt that difference in concentration of a given solute between two compartments linked by an exchange system is the best measure of its functional integrity.

It has been suggested that the concentration of lactate in fetal blood, or its gradient between mother and fetus, is a good indicator of fetal oxygenation. During acute asphyxia, concentration of lactate in the fetal monkey rises approximately by one mM/Lx min. Following reoxygenation the level remains elevated for a prolonged period, not unlike that of the excess hydrogen ion. The principal shortcoming of lactate as indicator of fetal well-being during labor and delivery is its time-consuming and complex determination. Although enzyme methods have substantially improved the measurement of this anion, the technics do not compare favorably with those available for the measurement of hydrogen ion concentration. It should also be pointed out that conceptually there is little advantage in choosing lactic concentration over excess hydrogen ion as indicator of fetal circumstance.

The measurement of carbon dioxide tension in fetal blood is readily performed on minute samples either by direct technic or by calculation from pH values of the blood sample equilibrated at two different tensions of CO_2. As can be seen in Figure 1, P_{CO_2} rises rapidly during asphyxia. Similar to hydrogen ion concentration, the ratio between rate of change and the accuracy of measurement is good. Paradoxically, the high time resolution of the variable is a drawback when intermittent evaluation of the condition is available, as is the case with the human fetus. The rapidity with which CO_2 can be cleared following asphyxia in the newborn is demonstrated in the same figure. If sampling of blood does not coincide with the episode of oxygen deprivation, a serious disturbance can be readily overlooked.

Similar limitations also pertain to the measurement of O_2 tension. In general, the tension of oxygen of blood obtained from the periphery is an unsatisfactory indicator of oxygen availability to vital organs. Even under optimal circumstances it is difficult to prevent oxygen uptake by hemoglobin during the collection period, particularly when the P_{O_2} of the peripheral blood sample is low. Cutaneous vasoconstriction or stasis produced by compression of fetal scalp during sampling may also yield specimens that reflect poorly the composition of blood in the central circulation. Furthermore, it is not possible to predict oxygen content of blood from P_{O_2} alone, because the affinity of hemoglobin for oxygen is markedly influenced by hydrogen ion concentration. Thus, in the presence of marked acidosis a nearly normal tension of oxygen may be present, while the oxygen content of blood is low. Since oxygen content rather than P_{O_2} is the principal determinant of oxygen availability of the tissues, this may lead to underestimation of fetal jeopardy.

A number of other constituents of fetal blood have been suggested to reflect changes in fetal oxygenation, such as osmolality and the concentration of red cells and glucose. Although a strong association may exist between these variables and oxygen availability to the tissues, they seem to offer no advantage over hydrogen ion as indicator of the adequacy of gaseous exchange across the placenta. It is probable, however, that in the future these tests will constitute an integral part in the biochemical assessment of the intrauterine patient.

contributors

Karlis Adamsons, M.D., Ph.D.
Professor, Obstetrics and Gynecology, College of Physicians and Surgeons, Columbia University, New York

Tom P. Barden, M.D.
Associate Professor, Obstetrics and Gynecology, University of Cincinnati College of Medicine

Donald H. Barron, M.D.
Professor, Physiology, Yale University School of Medicine, New Haven

Roberto Caldeyro-Barcia, M.D.
Professor and Chairman, Physiopathology, Universidad de la Républica, Montevideo, Uruguay

Arpad I. Csapo, M.D.
Professor, Obstetrics and Gynecology, Washington University School of Medicine, St. Louis

Emanuel A. Friedman, M.D., Sc.D.
Professor and Chairman, Obstetrics and Gynecology, University of Chicago Medical School

L. Gulin, M.D.
Montevideo, Uruguay

Charles H. Hendricks, M.D.
Professor, Obstetrics and Gynecology, Case Western Reserve University

contributors

School of Medicine, Cleveland (now Professor and Chairman, Obstetrics and Gynecology, University of North Carolina, Chapel Hill)

Edward H. Hon, *M.D.*
Professor, Obstetrics and Gynecology, Yale University School of Medicine, New Haven

A. A. Ibarra-Polo, *M.D.*
Montevideo, Uruguay

C. Méndez-Bauer, *M.D.*
Chief of Obstetrical Physiology, Pathophysiology, Faculty of Medicine, Montevideo, Uruguay

Kenneth R. Niswander, *M.D.*
Associate Professor, Obstetrics and Gynecology, State University of New York at Buffalo (now Professor and Chairman, Obstetrics and Gynecology, University of California at Davis)

J. J. Poseiro, *M.D.*
Associate Professor, Obstetrics and Gynecology, Faculty of Medicine, Montevideo, Uruguay

Elizabeth M. Ramsey, *M.D., Sc.D.*
Staff Member, Department of Embryology, Carnegie Institution of Washington, Baltimore

Richard W. Stander, *M.D.*
Associate Professor, Obstetrics and Gynecology, Indiana University Medical Center, Indianapolis (now Professor and Chairman, Obstetrics, University of Cincinnati College of Medicine)

Emil Witschi, *M.D., Ph.D.*
Senior Scientist Population Council, Rockefeller University, New York

Carl Wood, *M.D.*
Professor and Chairman, Obstetrics and Gynecology, Monash University, Queen Victoria Hospital, Melbourne, Victoria, Australia

index

Acceleration pattern, 94
Acetylcholine, 211, 213
Adrenergic mechanisms, in myometrial control, 75-89
Aldosterone, 212
Alphaprodine, 206
Amniocentesis, 79
Analgesia, 206-207
    ethyl alcohol, 214
Androgens, in male gonaduct development, 30
Anesthesia
    gaseous, 208-09
    blocking agents, 209
    spinal, 209
        pudendal, 210
    caudal, 210
    epidural, 210
Antispasmodics, 211
Apgar score, 91, 101, 184
    and fetal heart rate, 130-32, 199, 228
    and Type II dips, 135-47
    and fetal pH, 187
    and fetal pCO_2, 190
    and fetal pO_2, 191
    and intrauterine effects, 221
    and marginal sinus rupture, 230
    and maternal asthma, 230
    and maternal heart disease, 230

Asphyxia, intrauterine, 217-31
Asthma, maternal, and Apgar score, 230
Atropine, 212
Autonomic agents, 212-14
Basal fetal heart rate *see also* Fetal heart rate
    and Apgar score, 130-32
    definition, 130
    during labor, 130
    demarcation value, 131
    in prognosis of depression, 133-34
    post-atropine, and Apgar score, 134-35
Base, in fetus
    measurement, 183
    reduction, 184
        and maternal base, 187-90
Base ($\triangle$) deficit, 188
Baseline *see* Basal fetal heart rate
Biochemical measurement, ultramicro method, 183
Blood pressure, maternal, 79
Bohr effect, 188
Bradycardia, 94
Bradykinin, 214

Carbachol, 213
Carbon dioxide, transport from fetus, 184; tension in fetus, 240

245

prenatal life

Catecholamine, 86-89
Cerebral palsy, 217-18
Cerebrospinal fluid pressure, intrathecal medication contraindication, 70
Cesarean section, in fetal distress, 106
CSFP *see* Cerebrospinal fluid pressure
Chlordiazepoxide, 208
Chloride, measurement in fetus, 183
Chloroform, 209
Chlorpromazine, 207
Circulation
 umbilical
  development, 118
 uteroplacental
  development, 118
  oxygen level, 122
  estrogen and progesterone in, 125
  oxygen level, 122
Collaborative Project for the Study of Cerebral Palsy, 219
"Continuum of reproductive wastage," 218
Corpus luteum, regression, 180
Cyclopropane, 209

d-tubocurarine, 210
Decamethonium, 210
Deceleration pattern, 94
 and acute uteroplacental insufficiency, 96-98
 and umbilical cord occlusion, 98
 timing, 94
Decidua formation, human vs. rhesus, 38
Delta (Δ)
 base deficit, 188
Demerol
 affecting Type II dips, 137
 affecting sum of amplitudes, 141
Dibenamine, 212
Dibenzyline, 212
Dichloroisoproterenol, 76
Digoxin, 214
Dihydroergotamine, 212
Dimenhydrinate, 208
"Double isotope" technique, 167-69
Dramamine *see* Dimenhydrinate

Epinephrine, 75, 81, 213
 reversal, 86
Ergonovine, 212
Estradiol, 211
Estriol, 26
Estrogens
 In female gonaduct development, 30
 use in labor, 211
Ethyl, 209
Ethylene, 208

FHR *see* Fetal heart rate
FIP *see* Fetal/Placental ratio
Fetal acidosis, 98
Fetal blood
 gases, as fetal base, 189
 oxygen tension, 110
Fetal circumstance, indicators, 235-41
Fetal distress, 91-107
 diagnosis, 91-104, 129, 147-49
 deceleration pattern in, 94
 biochemical studies, 101-04
 indications from fetal heart rate patterns, 104
 treatment, 105-07
Fetal heart rate *see also* Basal fetal heart rate
 and Apgar score, 199, 228
 and placental insufficiency, 201
 and umbilical cord compression, 200
 auscultation
  correct method, 149
  incorrect method, 150
 baseline level, 93
 falls, 143
 in fetal well-being, 92
Fetal hypoglycemia
 and abruptio placenta, 194
 and growth retardation, 194
 and maternal hypoglycemia, 194
 and preeclampsia, 194
Fetal pH, and fetal well-being, 104
Fetal/placenta ratio, 166-67
Fetal scalp, temperature, 198
 and tachycardia, 198
Fetal scalp blood
 and Apgar score, 101
 biochemical analysis, 101
 measurement, 183-84

index

Fetal scalp electrodes, 197-98
Fetus
  environment, 104-26
  effects on uterus, 156
  anomalies, 160
  studies, 183-201
  diagnostic tests, 201
  biochemical analysis, 236-38
  biophysical analysis, 236
  central nervous system, and asphyxia, 236
Fornices, 25

Gallamine, 210
Genital ducts *see* Gonaducts
Glucose
  measurement, 183
  in fetus, 192-94
  and hypoxia, 193
  concentrations during labor, 193
  transport, and oxygen, 193
  and insulin level, 196
Gonadotrophins, 212
Gonaducts
  sexual differentiation, 12, 16-25
    male, 16
    female, 18-19
    endocrine factors, 30
  development
    progression, 19-30
    sources, 11
  estrogen responsiveness, 25-28

Haematocrit, measurement in fetus, 183
Haemoglobin, measurement in fetus, 183
Halothane, 209
Heart disease, maternal
  and Apgar scores, 230
  and perinatality, 228-30
Henderson-Hasselbach equation, 188
Heroin, 206
Histamine, 214
Hormones, 211
Hydrogen ion, in asphyxia, 238-40
Hymen, 28
Hypotension, and uterine contractions, 65-69
Hypoxia, 192
  intrauterine, 217-231

Imidazoline, 76, 213
Implantation, human vs. rhesus, 38
Inderal *see* Propranolol
Insulin, measurement in fetus, 183
Intracranial pressure, and deceleration pattern, 96
Intrauterine pressure, 79
  and deceleration pattern, 96
Isoproterenol, 75, 82
Isoxsuprine, 211

Ketonemia, 196
Ketones, 196-97
Ketonuria, 196

Labor
  abnormal patterns, 88
  initiation, 180-81
  drug effects, 205-15
Lactate, indicating fetal oxygenation, 240
Lateral position vs. supine position, 59
Levallorphan, 207
Librium *see* Chlordiazepoxide
Lidocaine, 209
Luteal function, suspension, 162
Luteoplacental shift, 167-69

Magnesium sulphate, 214
Marginal sinus, rupture, and Apgar score, 230
Meperidine, 206
Mesonephric ducts, 11-12 *passim*
Methadone, 206
Methylergonovine, 212
Morphine, 206
Myocardium, and asphyxia, 236
Myometrium
  and receptor sites, 77
  in vitro studies, 77-78
  in vivo studies, 79-80
  dysfunction
    progesterone therapy, 160
  function
    and adrenergic mechanisms, 75-89
    in intact gravid human uterus, 84-86

N-allyl normorphine, 207

prenatal life

Neostigmine, 213
Neurologic damage, 217
 from asphyxia, 218
 from hypoxia, 218
Nicotine, 213
Nitrous oxide, 208
Norepinephrine, 75, 80, 84, 213

Ovariectomy
 and placental hypertrophy, 165-66
 and placental retention, 166-67
 and premature delivery, 166-67
 timing, 162
"Ovariectomy syndrome"
 and labor, 155-81
 definition, 158 *passim*
Oviducts, 11-16, 19
 development, 12
Oxygen
 pressure
  fetal adaptation, 112
  tension in fetus, 241
Oxymorphone, 206
Oxytocin, 84, 211

Paraldehyde, 207
pCO_2, in fetus
 and Apgar score, 190
 increase, 184, 190
 measurement, 183
Penis, 17
Pentobarbital sodium, 207
Perinatality
 and maternal shock, 224
 and maternal heart disease, 228-30
Perphenazine, 208
pH, in fetus
 measurement, 183
 reduction
  and blood base, 184
  and pCO_2, 184
 and Apgar score, 187
Phenazocine, 206
Phentolamine, 80, 84, 213
Phosphate, measurement in fetus, 183
Physostigmine, 213
Pilocarpine, 213
Piperoxan, 213

Placenta
 circulation, 37-52
  effects of myometrial contraction, human vs. rhesus, 40-43
  arterial inflow, maternal, 44
   during uterine contraction, 46
  in fetus, 46
  and acid transport, 186
  radioangiography, human vs. rhesus, 43-44
 diffusing capacity for oxygen, 124
 diffusing capacity for urea, 123
 dislocation, 176
 endocrine hypertrophy, 165
  treatment, 172-76
 insufficiency
  and fetal heart rate, 201
 limits of size, 115
 local effect on uterus, 176-77
  electrophysiological documentation, 176-79
 retention
  an ovariectomy, 166-67
 venous drainage, maternal, 45
  during uterine contraction, 46
pO_2, in fetus
 measurement, 183
 and Apgar score, 191
 and oxygen inhalation, 192
 of fetal scalp tissue
  and oxygen inhalation, 197
  measurement, 197
Poseiro effect, 65
Potassium, measurement in fetus, 183
Pregnancy
 maintenance, rabbit vs. man, 175
 pharmacologic studies, 79-80
Priscoline *see* Tolazoline
Procaine, 209
Prochlorperazine, 208
Progesterone
 effect on fetal/placental ratio, 167
 therapy
  decremental, 173-76
  in ovariectomy, 171-72
 "depot" effect, 175
 use in labor, 212
Promazine, 207

index

Promethazine, 207
Propranolol, 76, 80, 84, 86
Prostatic urethra, 16
Prosympal, 213
"Pump-priming" effect, 69

Receptor sites, alpha and beta, 75
  effects of sympathomimetic agents on, 76
Regitine *see* Phentolamine
Relaxin, 212
Rhesus monkey, reliability as a human model, 38-44

Scopolamine, 212
Sedatives, 206
Shock
  maternal, 231
  and perinatality, 224
Sodium, measurement in fetus, 183
Succinylcholine, 211
"Suction electrode," 177
Supine position vs. lateral position, 59
Suxamethonium, 210
Syntocinon *see* Oxytocin

Tachycardia, 94
  in fetus
    diagnosis and prognosis, 129-151
    scalp temperature, 198
Tetracaine, 209
Tetraethylammonium ion, 213
Thiopental sodium, 207
Tolazoline, 213
Tranquilizers, 207
Tribromoethanol, 207
Trichlorethylene, 209
Trophoblast, human vs. rhesus, 38
Tubes, development, 28-29
Type II dips, 129-31
  definition, 129
  in fetal distress, 147
  by "total number"
    and Apgar score, 136
  and Apgar score, 135-47
  by "percentage of uterine contractions"

Type II dips—*Cont.*
    and Apgar score, 138
  by sum of amplitudes
    and Apgar score, 139-41
  by mean amplitude
    and Apgar score, 141-42

Umbilical cord, compression, and fetal heart rate, 200
Urogenital sinus, 11, 16-17 *passim*
Uterine contractions
  and hypotension, 65-69
  effects, 55-73
    on circulation, 55
    on blood pressure, 57
    on cardiac output, 57
    on heart rate, 57
    on intrathoracic pressure, 57
    on lactic acid levels, 62
    on abdominal pressure, 69
    on cerebrospinal fluid pressure, 70-72
  effects, mechanical, on circulation, 63-65
  timing, 149
Uterine glands, development, 29-30
Uteroplacental vessels, transformation, 40
Uterovaginal rudiment, 31-32
Uterus
  development and differentiation, 11-33
  epithelium, 23
    mucification, 33
  blood flow, 119
  function, laboratory studies, 156
  oxygen consumption, 119
  sensitivity, 206
Utricle, 16

V/P ratio, 169-70
Vagina
  development, 22, 31
  estrogen sensitivity, 26
  epithelium, keratinization, 33
Vagina bud, formation, 18-22, 32-33
Vestibulum, formation, 19

Harold C. Mack is professor of Obstetrics and Gynecology and former chairman of the department at Wayne State University School of Medicine. He received the B.A. and M.D. degrees from the University of Michigan.

The manuscript was prepared for the printer by Robert H. Tennenhouse and Linda Grant. The book was designed by Donald Ross. The text typeface is Linotype Palatino designed by Hermann Zapf. The display face is Optima also designed by Hermann Zapf.

The book is printed on S. D. Warren's Olde Style Antique and bound in Columbia Mills' Bayside Linen. Manufactured in the United States of America.